As they have done for millennia: bison grazing the prairies (Badlands National Park, SD)

For Willy Dietzmann, agricultural extension agent and my grandpa.
He was full of love for his granddaughter, passionate about farming
and he knew where the best meat and sausages were to be had. I
still miss him.

The Sustainable Meat Challenge

How to graze cattle, slaughter humanely and stay profitable

Written by: Marianne Landzettel

Photography: Martin Kunz

The Sustainable Meat Challenge
How to graze cattle, slaughter humanely and stay profitable

Marianne Landzettel

ISBN: 978-3-922845-58-4
Published by: World University Service, Deutsches Komitee e. V.,
Goebenstraße 35, 65195 Wiesbaden

All photography © Martin Kunz
Book design by www.SiPat.co.uk.

Table of contents

Foreword by Patrick Holden 6

Part 1: Setting the scene: Why this book and why now?

1. Why we need animal agriculture. 10
2. Making the case for small abattoirs in the UK 38
3. Humane, local and profitable – how it can be done 47

Part 2: New rules – solutions from across the Channel

4. Regulated down to the second 55
5. A good life and death within the herd 60
6. When small abattoirs work best. 72
7. Butcher training, humane slaughter and meat quality 101

Part 3: Across the Atlantic: prairies and 'niche meat'

8. Prairies, bison and the Dust Bowl. 119
9. MSUs and abattoirs: big and sophisticated or small
   and simple 132
10. Processing meat efficiently and building a local
    food system 151
11. Skills needed: meat cutting, management and marketing 181

Part 4: Humane, local and profitable meat in the UK

12. Lessons from the EU and the US and how to implement
    them here 197
13. In a time of change in agriculture, local slaughter
    creates opportunities 213

Foreword

by Patrick Holden

I was delighted to be invited to write the foreword for this important book covering one of the most crucial, yet often forgotten, elements of our meat supply chains.

My organisation, the Sustainable Food Trust, has been heading up a campaign to save the UK's small abattoirs for the past six years. They provide an important service without which it would be impossible for any of us to buy locally produced meat.

But a big part of the reason why I feel passionately about this issue is because I am personally involved. I am a livestock farmer, producing pigs and ruby veal beef. Our animals are slaughtered and processed at an abattoir only six miles from our farm. When you participate in this process - of ending the life of an animal - you realise that the story of its slaughter really matters.

In an ideal world, the animal would die on the farm. In the UK, this is currently illegal if the meat is to be sold to customers. The next best option is to take a very short drive to the abattoir, with minimum stress and gentle handling right up to the moment where the animal is killed.

This compassionate approach has become near impossible for the vast majority of livestock farmers in this country. As recently as 30 or 40 years ago you could find an abattoir within a few miles of your farm, but more than 85% of red meat abattoirs have closed in the past 50 years. There are now only 64 smaller abattoirs left in England and they are still closing at an alarming rate.

I'm becoming increasingly convinced that hundreds of thousands, maybe millions of the young and not so young people who are becoming vegan, or at least switching to a mainly plant-based diet, are doing so because they are deeply uneasy about the story behind supermarket meat and dairy products.

And who can blame them, many such animal products are from animals that are farmed intensively - they are treated as units of production and kept in confined spaces, some never even seeing the light of day. They then suffer stress in their final journey. Supermarkets have a policy, for reasons of economy of scale, of only using one or two abattoirs per species, as a direct result of which animals must travel long distances to slaughter, before disappearing into a vast and anonymous food processing system and ending up on supermarket shelves.

One might argue that by centralising and industrialising the slaughter of the animals they source, supermarkets are losing their most critical future market - the next generation. Yet to my knowledge, few food companies or retailers seem aware of the consequences of such a course of action.

Interestingly, quite a number of young people visiting our farm who come as vegetarians leave deciding that they would eat meat if they knew it had a good story behind it.

That is why this new book, by Marianne Landzettel is so timely and welcome.

This book shines a light on people doing things differently – those who have had the patience, imagination and determination to find ways to ensure the animals in their care have the best life but also the best possible death.

It shows that what is at stake matters a very great deal – it is about the ability to buy good meat with a humane story behind it from local farmers. It shows how local and on-farm slaughtering and butchery can revitalise rural communities and enable the transition to regenerative food systems. Such systems harness the power of ruminant animals to turn grass into highly nutritious food that we can eat, whilst at the same time, restoring the health of our soil, tackling climate change and reversing biodiversity decline.

We owe a great deal to Marianne Landzettel for her persistence in this research and determination to share her learnings. While focusing on Europe and the United States, the book explores how some of these examples could be replicated here in the UK, where we urgently need to revitalise our local abattoir sector.

For society as a whole, this book calls into question the future story we need behind our food. Are we really happy to be participants in an industrialised system which gives no heed to the welfare of the animals involved or the provenance of the meat we eat? We could all learn from this important research and think differently about the future food systems we wish to see.

Part 1

Setting the scene: Why this book and why now?

Chapter 1

Why we need animal agriculture

When I set out to write about abattoirs and mobile slaughter units, I knew that I would see slaughtermen and butchers at work, that I would witness animals dying. What I didn't know was how much I would have to think about death.

In western societies, not much thought is given to death. Since most kids grow up in nuclear families, they rarely witness death – nursing homes or hospitals are where death happens to people. Children may have to cope with a pet dying, but they are unlikely to see a pig being slaughtered or even a chicken.

Industrial agriculture has made it possible for us to eat meat and consume dairy without ever giving a thought to where it came from and how it was produced. Climate change, the COVID-19 pandemic – which led more people to once again buy directly from farms – and discussions about sustainable production may have changed attitudes slightly, but we surely don't think about slaughter when we take steaks, pork chops or sausages out of a supermarket chiller cabinet where they sit, neatly packed and portioned.

The convenience of this type of supermarket shopping wouldn't be possible without industrial agriculture. Yes, we do know that farmers provide the food on supermarket shelves, but how it gets there is usually not something we need to think about – this is not just convenient for the industry, it's essential. Would we buy beef if we thought about cattle spending months in grassless feedlots?

Would we have a pork chop if we knew it came from stressed tail biting pigs? A number of NGOs, animal rights campaigners and documentary filmmakers have thrown a light on untenable conditions in animal agriculture and raised health concerns over pesticide and antibiotic use. But what about slaughter?

"My animals have one bad day in their lives, and that's their last" is what many farmers will tell you. And that claim is likely true. Most farmers don't adhere to animal welfare standards because it's the law or because they get a premium, but because they care for their animals and treating them well is the right thing to do.

But what about that last day? What happens on that day is a matter for the people who transport the livestock and the slaughtermen at the abattoir. We, as consumers, don't ask what happens on that day, nor do many farmers. But some do, changing the way they work, and fighting for regulatory changes: as I started to research small abattoirs I met butchers and farmers for whom this last day, the last hours, the final moments in an animal's life matter.

Livestock is part of any diverse farming enterprise and that means animals will have to be slaughtered at some point. These concerned farmers and butchers are defining what *humane* slaughter looks like. They want their animals to have a good life – and a good death.

We cannot change what we don't know. That's why we need to look at slaughter, who does it and how it's done. Humane slaughter is possible – on farm and in small abattoirs.

But do we really need livestock on farms? Isn't a diversified small farm with some arable land, beef or dairy cattle, possibly chicken, too, an utterly outdated model?

Over the past decades, specialisation in farming has increased productivity and, depending on the size of the enterprise, profitability. Dairy cows, pigs and poultry can be housed and managed efficiently, beef cattle may spend a short time on pasture before it's finished in a feedlot. Arable farms don't need animals, the available land can be used for growing food and feed crops such as wheat, maize, soy, sugar beets, oil seeds. The distinction now is between livestock and arable farms, a specialisation that has been made possible through agrochemicals, chemical fertilisers and pesticides.

A worm count is a good indicator for soil quality.

Specialisation, maximising production and increasing profitability through expansion – more animals, more land – is at the core of the industrial farming model. Proponents will argue that only through industrial agriculture farmers are able to produce enough cheap food to feed a growing world population. Critics point at the environmental costs of industrial agriculture – polluted waterways and dead zones in the oceans, air pollution, antibiotic resistance, 'superweeds' that are resistant to numerous types of chemicals, erosion and dead soils. For the purpose of this book I want to focus on soil: fertile soil is the basis for all food production. Animals – the livestock above ground as well as the "livestock" below ground - play a key role in building and maintaining soil fertility. And their habitats are not just connected, neither thrives without the other.

Before we take a look at why the livestock above ground, cattle in particular, are so important not just for permanent grasslands but for arable farms, too, we need to start with the 'livestock *underground*', the myriad of microbes, earth worms, insects and single cell organisms that populate 'good' soil. Why? The short and concise explanation comes from Gabe Brown, the farmer and author from North Dakota. He sums up his transition from being a commodity farmer to becoming a trailblazer for regenerative agriculture in just three word: "Dirt to Soil".

For the first years after he purchased his farm he worked the same way his neighbours did: "I continued to farm conventionally using tillage, fertilizers, and herbicides to grow small grains, as my in-laws had been doing. I simply did not know any other way; it was what I had learned in college and from (his father-in-law) Bill"[1]. Soil was commonly referred to as 'dirt', useful to accommodate the roots and keep the plants upright. All plants needed to grow was N-P-K, available from various agricultural supply companies. It seemed all the farmer had to do was to get enough of it onto the field at the right time.

In the late 1990s Brown went through what he calls 'the disaster years' during which he nearly lost the farm. "Today, I tell people that those four years of crop failure were hell to go through, but they turned out to be the best thing that could have happened to us, because they forced us to think outside the box. to not be afraid of failure, and to work with nature instead of against her. They sent me on this journey of regenerative agriculture"[2].

There may not be a universally agreed upon definition of what 'regenerative agriculture' is. On the one hand, food and agricultural companies have tried to appropriate the term in recent years, on the other hand organic farmers say that they've been using regenerative practices all along and if anything, the term should be agroecology. But an emphasis on 'soil health' can be found in any definition.

The existence of soil life and its importance for soil fertility was recognised early on by pioneers such as the British biologist Sir Albert Howard or the farmer and founder of the Soil Association, Lady Eve Balfour. In the US, the agronomist William Albrecht was the eminent authority on soil health. He became the chairman of the Department of Soils at the University of Missouri and not only made the connection between soil health, food quality and the health of animals, in later life he also criticised the reliance on chemical on NPK fertiliser for the depletion of soils.

A definition of the term 'healthy soil' needs to be comprehensive, says agroecologist Nicole Masters: "Any definition of soil health must include profitability as well as crop quality. Definitions around soil health must consider, too, reductions in greenhouse gases emissions, buffers to toxins and resilience to climatic pressures. Healthy soils have functional water, carbon, decomposition and mineral cycles. These soils are vibrant and alive with full structural integrity, able to withstand heat, dry, floods and rebound quickly after disturbance"[3].

Soil does not get better than this.

Soil science is a fast developing discipline and our knowledge of the soil ecosystem is still in its infancy. "When I was an undergrad student we thought we knew about 10% of soil organisms. Today we think it's maybe 1%"[4], US soil scientist Kris Nichols told me when I interviewed her in 2018. "There can be billions of organisms in a handful of soil, invisible to the naked eye. (...) In healthy functional soil, there are 6 major livestock classes: viruses, bacteria and archaea, fungi, protists, nematodes, micro-arthropods and micro-animals", says Masters.[5]

"Fostering life is the key to transforming *dirt* into *soil*"[6], says Gabe Brown. Agrochemicals do the opposite, they transform soil into dirt, as Brown learnt when Kris Nichols visited his farm. "She explained that synthetic fertilizer is detrimental to mycorrhizal fungi. By applying fertilizer, I was actually hurting my soil, not helping it. Synthetic fertilizer interrupts the relationship between microbes and plant roots because the fertilizer gives plants 'free' nutrients, so they don't need to trade carbon for nutrients from microbes. When that happens, the plants keep a lot of that carbon

You can smell good soil. Gail Fuller, regenerative farming pioneer, KS.

for themselves, which means the microbes don't get enough food to grow and reproduce, and their populations suffers. Mycorrhizal fungi efficiently acquire minerals for plants in exchange for carbon, but if the fungi do not receive carbon from the plants, they can't acquire minerals, and then plants can't access those minerals. Also, synthetics supply only a limited type of nutrients, not the full range that plants need"[7].

Brown decided to put Nichols' advice to the test. In 2004 he split several fields in half, one half received fertiliser, though a smaller dose than recommended, on the other half he spread no fertiliser at all. "For four years in a row, the crop yield of the unfertilized half of these test fields were *equal to or greater* than the fertilized half! I also noticed a dramatic improvement in the health of our soils once

I removed synthetic fertilizer. The soil was much more aggregated. It really was like chocolate cake! This aggregation meant water infiltration had improved significantly"[8].

This is rather bad news for agrichemical companies. Synthetic fertilizer is one important part of their business model, the other one is 'crop protection', as the industry likes to call it. 'Pesticides' is the more appropriate term because the products do what it says on the canister: they kill living organisms, the absolute minority of which are actual pests. This is how Nicole Masters puts it: "Many of the crude, broad-brush chemical controls have set agricultural systems up for the proliferation of pests and diseases. In a chemical arms race, it's the insect pests who are winning the war. For every 1 pest species, there may be as many as 1,700 non-pest insects who have become the unintended casualties of this war. Insects provide a multitude of ecosystem benefits from pollination, nutrient cycling, decomposition and fueling the foodweb. The impacts from a looming 'insectaggedon', the collapse of insect species, is broad ranging, far-reaching and potentially catastrophic. Although non-target species, like bee and butterfly populations are collapsing, the crop pest species are flourishing. There are now over 550 insect species resistant to pesticides, including insects that have evolved to consume the *Bacillus thuringiensis* (BT) toxin contained in engineered corn, cotton, soy and potatoes. Despite and increasing complexity of chemical controls, pests still consume 18-20% of the global crop and are becoming increasingly resistant to the controls. (...) A 2018 study in the US corn belt comparing regenerative farms to conventional farms using insecticides, found ten times more insect pests in the conventional. Yup, you read that right, where farmers were applying their full arsenal of insecticides, genetically engineered, plants and seed treatments, there were 10 times more insect pests"[9].

Industrial agriculture.

Masters continues: "Pesticides are the most inefficient of all the agrichemicals. It is estimated that at best 1% of these chemicals reaches their target sites, as nearly all is lost to run-off, spray drift or degraded in sunlight. In the case of neonics, only a tenth of the seed treatment is taken up by the plant, leaving the remaining 90% to impact on non-target species in soil, dust and waterways"[10].

Pesticides may be inefficient at killing pests, but they are extremely efficient at killing everything else. Whether they are applied above ground or below as seed treatment, in the end most pesticides will end up in the soil, either through run-off, water infiltration or as dust. And there they will wreak havoc: a fungicide is designed to kill fungi and it does not discriminate between leaf rust above ground and mycorrhiza below ground. Pesticides not only decimate the 'livestock underground' such as microorganisms, worms, fungi and insects, but they destroy the whole soil microbiome and with it soil structure. Soil becomes dirt. On the other hand: healthy soil with a diverse community of microorganisms will protect plants against disease.

The top soil is almost gone.

Depleted soil: rain turns it quickly into mud because the water infiltration is low.

Everything hinges on photosynthesis, says Nicole Masters. The better a plant is able to photosynthesise, the more sugars (measured in Brix), proteins and lipids it can produce and the healthier the plant is. In his talks and training sessions, farmer and researcher John Kempf explains the connection between optimal photosynthesis and plant health through the plant health pyramid. "As soils and crops transition with regenerative farming practices, they pass through stages of increasingly better health. The progression to better health restores the natural and biological abilities of the plant and soil system. During this process, plants will demonstrate increasing immunity to soil and airborne pathogens, better resistance to insects, improved production of lipids leading to stronger cell membranes for tastier fruit with better shelf life, and more"[11].

In her book 'For the Love of Soil', Masters relates the results of data taken from 22 regenerative properties in New Zealand: the higher the Brix levels the better the plants photosynthesise and the lower the pest pressure. "As a direct result from their soil and pasture health programmes, these farmers saw a lift in Brix levels, while their insect pest pressure dropped. Insects are not after every blade of grass or every apple. They're looking to clean up garbage! Establishing and maintaining a diverse microbial community and building soil organic matter, is essential to reduce environmental stressors, such as drought and temperature fluctuations Microbes can reduce nutritional stresses by ensuring plants receive minerals, at the rate, time and form they require. Soil and leaf microbes provide a range of beneficial services to boost plant health[12]".

A small group of US farmers pioneered regenerative agriculture together with soil scientists like Ray Archuleta, Kris Nichols and Christine Jones. After 2000, just a few years after the introduction of glyphosate resistant GM corn and soy, more and more farmers were at their wit's end because they spent more and more money on GM seeds, fertiliser and herbicides - for ever lower returns. At the same time, weather patterns began to change dramatically, the eastern part of the 'corn belt' was often too wet, the western part and the High Plains too dry. In 2012, 80% of the US were hit by a severe drought. These 'disaster years' forced many of these farmers to work in a different way.

Gabe Brown was among the farmers who shared their story at the No-Till on the Plains conference which, in the following years, became a meeting point and a refuge for farmers who wanted to work differently, with nature instead of against it. During the research for my book 'Regenerative Agriculture: Farming with Benefits', I had the good fortune to meet several of them on their farms and see how the soil quality improved as the soil organic matter increased. Early into my visit the farmer would grab a spade, push it through the layer of grass or cover crops and expose crumbly, 'chocolate cake' soil, with worms hanging out on the side.

Soil health and abundant soil life were certainly at the centre of everything these farmers did. But why and how does it all work? How can progress be measured? Which practices work best and under what conditions?

Cynthia Daley is the founder and director of the Center for Regenerative Agriculture and Resilient Systems at the California State University in Chico. We met in Chico in early November 2018. She wished I were able to come a week later, Daley had told me in an email, Gabe Brown, Ray Archuleta and Dave Brandt, a farmer from Ohio who was one of the first to work with cover crops and underseeding corn, were to hold a seminar in Chico. As it turned out that seminar never happened, on November 8th, the Camp Fire broke out, to date the deadliest wildfire in the history of California.

"There is no normal anymore, all we have is abnormal", Daley told me when we met the following day in a trailer on the Chico University farm. To deal with the climate crisis, "we need all hands on deck". And as the pioneers of regenerative agriculture farming in different locations across the country have tried and tested different farming methods it was now up to the scientists to go to the farms and study why these systems work so well.

The Center for Regenerative Agriculture in Chico does research as well as training, there are numerous online resources and mentor farmers give practical advice. The website defines regenerative agriculture as "an approach to farm and ranch management that aims to reverse climate change through practices that restore degraded soils. By rebuilding soil organic matter and soil biodiversity we significantly increase the amount of carbon that can be drawn down from the atmosphere while greatly improving soil fertility and the water cycle"[13].

Fertility building
herbal leys on
an arable field.

A new area is
fenced in for the
cows to move in.

The approach is summarised in five essential principles[14]:

+ Maintain soil cover throughout the year.
+ Minimize soil disturbance.
+ Maintain living roots throughout the year.
+ Support diversity of vegetation community.
+ Incorporate grazing animals.

Which finally brings us to the 'livestock above ground'. The last principle, incorporating grazing animals, makes the first four work in the most effective way: maintaining soil cover and living roots throughout the year means working with cover crops. A multi species cover crop mix and long rotations including herbal leys support the diversity of vegetation community. And minimizing soil disturbance means ploughing as little as possible or not at all.

In a conventional no-till system farmers will use herbicides such as glyphosate to clear fields after harvest and eliminate weeds before seeding, a practice known as 'fall and spring burn'. The same goes for cover crops, which need to be 'terminated' by spraying them with herbicides or by ploughing. Both practices harm soil organisms and soil structure.

No glyphosate needed to 'terminate' the herbal leys. The cows enjoy the species rich mix and fertilise while they graze.

What are the alternatives? Cover crops can be flattened and shredded by using a roller-crimper. Or they can be grazed by livestock. The same goes for wheat and maize stalks or the remnants of soybean plants: cattle and sheep happily glean the fields. On organic and regenerative farms livestock has become a vital 'tool' – during their lifetime they are four-legged land managers.

Cattle provide vital eco services: on arable farms, on permanent pastures and in habitat preservation. Bovines should therefore be considered a keystone species.

ATTRA is a programme of the US National Center for Appropriate Technology (NCAT), and, according to their website, funded through a cooperative agreement with the United States Department of Agriculture's Rural Business - Cooperative Service. One of their publications, 'Integrating livestock and crops: improving soil, solving problems, increasing income',[15] points out that apart from the obvious financial benefits of livestock such as income from meat, milk and wool, ruminants in particular should be given credit for the many services they can provide on a farm. ATTRA lists:

"Livestock can:

+ Increase soil organic matter, which increases water-holding capacity.

+ Improve soil fertility, adding nitrogen, phosphorus, potassium, and many micro nutrients essential for plant growth.

+ Increase soil life and biodiversity, which improves soil functioning.

+ Reduce pest problems, including wheat saw fly and alfalfa weevil, by grazing during dormant season to remove pest habitat.

+ Remove weeds in fallow ground to not only reduce weed problems but also conserve soil moisture.

+ Reduce waste on the farm by grazing residue, consuming spoiled vegetables and fruits, or grazing damaged crops. This salvaging function enables income from crops even in a difficult year when crops are unsalable.

+ In orchards, animals consuming dropped fruits can greatly reduce pest problems, including plum curculio.

+ Grazing buffer zones, lanes between tree rows, and riparian edges can help maintain the landscape while making these areas productive parts of the farm.

+ Enable income during rotations that otherwise be strictly cover crops; turn cover crops into another cash crop by grazing, improving soil, and breaking pest cycles, while still producing a saleable product.

+ 'Close the loop' of farm cycles, making a more self-sufficient unit.

+ Produce other products for diversified income streams, better cash flow, less risk, and greater consumer interest.

+ Add animate life to the farm, increasing interest for the farm family and for visitors."

And: "The synergy of connecting crops and livestock can also apply to connecting crop and livestock producers: each component is stronger by being associated with the other".

In Britain, the authors of the Organic Farm Management Handbook 2023 have a similar assessment of the role of livestock on organic farms: "Livestock are integral to most organic farming systems, particularly those with permanent pasture and ley/arable rotations. The only exceptions are stockless horticultural and arable farms resulting from specific moral or practical decisions made by the farmer, such as vegan organic systems, or where there is an absence of farm infrastructure to support livestock farming. Ruminant livestock are able to utilise and provide a financial return to the leys with legumes and other herbage species that contribute to nitrogen fixation and the fertility-building phase of the rotation. They can utilise cellulose and hence energy from herbage that would not otherwise be available for human consumption and the return of livestock manures provides a targeted nutrient source for subsequent crops. Livestock thus play an important role in nutrient and energy cycling. They can also contribute to weed, pest and disease control through grazing and forage conservation. As with crop production, an emphasis on single species can lead to problems, including internal parasites or weed (e.g. bracken in permanent pasture). The mixing of species, such as sheep and cattle or sheep and poultry, can contribute to the control of parasites and improve grassland management"[16].

In a recent article in Farmers Weekly George Greed, a Devon farmer who converted to organic and grows population wheats, explained how integrating herbal leys and beef cattle gave the farm business a real boost: "(...) the biggest game changer was integrating the 125-cow suckler herd and Herbal leys into the arable rotation. The species-rich leys include: red and white clover, an array of grasses, chicory, plantain and birdsfoot trefoil. 'We previously kept the beef and arable separate – other than spreading manure – but grazing herbal leys transformed our soils in just two-and-a -half years. Soil nutrients are cycling and biology is thriving. We now see many more insects and bird species with groups of starlings following the cattle when bale grazing'"[17]

Much has been made of the difference between organic and regenerative farming. During my research for 'Farming with Benefits' I talked to both regenerative and organic farmers and asked them what they considered to be the main differences between the two systems. It boiled down to what they considered the 'measure of last resort' when it comes to weeds: while organic farmers per definition are not allowed to use chemical pesticides they will occasionally consider it necessary to plough, for weed control and, under specific conditions, to break up a hardpan. When they plough they will use a chisel plough rather than a mouldboard plough to keep soil disturbance to a minimum and avoid turning soil layers upside down.

For regenerative farmers the mantra usually was 'never ever plough', but many would consider herbicide use as a measure of last resort.

With almost all other farming practices being identical – from the emphasis on soil health and biodiversity to the use of cover crops, long rotations and the integration of livestock wherever possible – why would regenerative farmers not contemplate going organic? The answer is cost: most felt that organic certification and the necessary annual audits would cost them more than the additional income they could achieve through the organic premium paid for their produce. This had to do with a lack of infrastructure, for example the difficulty to access an organic certified abattoir, lack of market access and/ or lack of demand.

On its website, the Center for Regenerative Agriculture in Chico makes this argument for livestock integration: "Although raising livestock and crops together used to be the norm, in the United States farm production has shifted to increased specialization because of presumed management efficiency. Considerable research, however, has found that reintegrating animals into crop production systems yields considerable benefits in improved soil health, reduced risks associated with raising a single product, reductions in fertilizer input and animal feed costs, reduced labor and machinery costs, and increased carbon sequestration. Grazing cropland improves soil fertility by increasing soil microbial density and organic matter due to the addition of manure. It can also provide significant benefits for farmers who use cover crops and no-till methods as the animals can graze the cover crops while lightly integrating their manure into the soil with their hooves. Managed grazing and crop rotation techniques work best with this approach to avoid over-compaction of the soil"[18].

Whether farmers plant cover crops between cash crops to keep living roots in the ground year round or work with herbal leys which are in the ground for several years to help build soil fertility, both provide excellent cattle feed. Under resources, the Center for Regenerative Agriculture lists the "Grazing cover crops: a how-to guide" which specifies the benefits of integrating livestock in more detail: "Studies and field trials have shown that combining cattle grazing with cover crops, especially multi-species cover crops, accelerate the increase in SOM[19], which improves the soil's water holding capacity. Each one-percent increase in SOM results in approximately 27,000 more gallons of water holding capacity per acre (NRCS, 2013). A USDA-ARS field study showed cow/calf pairs grazed for 48 days +/- on cover crops of pearl millet, rye, rye-ryegrass and rye-crimson clover. This trial included plantings of full season cover crops following winter wheat. Grazing also has economic benefits by providing forage for cattle and other ruminants. In addition, grazing cover crops with livestock provides fertility for the next cash crop. Analysis performed by Grass Fed Insights, LLC shows that a 1,000 lb cow or steer produces approximately 0.25 lbs of Nitrogen (N), 0.15 lbs of Phosphorus (P), and 0.52 lbs of Potassium (K) per day in their manure and urine (unpublished data). By implementing adaptive grazing practices in cover crops, 100 cows averaging 1,000 lbs grazed across one acre in one day can deposit a total of 23N-15P-52K on that acre in a single day. If three grazings of each acre of cover crops can be achieved in a single cover crop season, then a total of 69N-45P-156K will have been applied to fertilize the subsequent cash crop. This does not take into consideration the additional contribution of the cover crop itself"[20].

Integrating livestock in arable systems is only part of the story. In grassland systems, keeping livestock isn't just a possibility, it is the only option. In his article 'The importance of grazing livestock for soil and food system resilience' author and farmer Richard Young writes: "Grassland covers 30 per cent of ice-free land and 70 percent of agricultural land globally. It plays a uniquely important role in food system sustainability, in addition to its well-understood roles in relation to nature conservation, water quality catchment management and the culture of some societies. Grass will grow on land that is too steep, too stony, too acidic, too wet and too poor for crop production. As such, grazed grasslands produce food from large areas of land that would otherwise be unproductive. Soils store four times more carbon than all trees and other life on the planet and tree times more than the atmosphere[21]."

The emphasis here is on *grazed* grassland. "Unlike trees and crops, over 60 million years grasses co-evolved with grazing animals in a unique way. Grazing triggers a growth impulse"[22], says the veterinarian Anita Idel who has written extensively about permanent grassland and the positive climate impact of grazing cattle. The growth impulse increases the rate of photosynthesis, the plant absorbs more CO_2 from the air into the chloroplasts, the part of the plant cell where photosynthesis happens. "Both, annual and permanent grasses share the same blueprint. A grazed blade of grass has been torn of at the top and keeps that tear-off edge while growing from its base and once more increasing in length[23]".

Cows rip off grass with their tongue. That triggers photosynthesis in the grass plants to go into overdrive.

Grazing leaves a ripped edge. It remains because grasses regrow from the bottom. Trees and other plants are different: when shoots and leaf buds get eaten they will not regrow.

Grasses invest the energy not only in the regrowth of the grass blades, the green part of the plant, but they also increase the root mass below ground. Grasses are the 'interface', this is where the livestock above ground and the underground livestock 'meet'. An increase in root mass leads to an increase in root exudates which feed soil organisms. But: "Grasses have to reduce their root mass if they are mowed or grazed too short. A part of the root mass is then not available for soil building"[24], writes Idel.

Not only does a large part of agricultural land globally consist of grassland which cannot be utilised for food production, by raising livestock, grazing permanent pastures brings environmental benefits. "If the grazing is controlled, then much of the carbon dioxide that the grass fixes by photosynthesis finishes up in the roots rather than the leaves and remains uneaten, and, as the roots die so the carbon content of the soil is increased, so that a well-grazed field can be a net carbon *sink* – the very thing the world needs. Thus, grazing livestock (in a controlled way) should be win-win"[25], says Colin Tudge, biologist and author of 'Meat: A Benign Extravagance'.

Short grass prairie in eastern Colorado. Grasses and cattle have co-evolved over millennia. For grasslands to thrive they need to be grazed and fertilised in the process.

Nevertheless, if you read newspaper headlines you would be forgiven for concluding, that ruminants, burping cows and sheep, are among the main culprits causing the climate crisis. Meat and dairy production not only have a large carbon footprint, so the argument, but through their burps ruminants also release a lot of methane.

Let's look at methane first. What the headlines don't say: in regard to emissions not all cows are equal. In their book about beef production and regenerative grazing, Grass-Fed Beef for a Post-Pandemic World, Ridge Shinn and Lynne Pledger write: "Many allegations about cattle's contribution to atmospheric methane may be true for conventionally raised cattle but are not true for 100% grass-fed beef cattle raised regeneratively. Bovines raised with regenerative methods eat higher-quality forage than cattle raised conventionally. Green leafy plants offer cattle better nutrition than corn and concentrates. Cattle digest higher-quality forages more quickly, reducing methane burps and lowering the amount of methane that the animal generates. Methanotrophic bacteria significantly reduce emissions because they live in pasture soil (among other places), and as cattle graze, these bacteria oxidize methane as their sole energy source. Of course, this beneficial process does not occur where cattle are housed in a feedlot – or when cattle are removed from the pasture environment and enclosed in stainless steel rooms to measure their methane output. Also, tillage, nitrogen fertilizers,

Cattle on mixed grass prairie in Nebraska.

and bare land destroy methanotrophic bacteria. Scientists are also learning more about how water vapour transpired from pasture plants creates an oxidation zone whereby hydroxyl radicals break down methane. Again, this process takes place in the *context of a pasture*, not in a feedlot or stainless steel laboratory box. Because grass-fed beef cattle are not confined, they drop their manure all over the pasture. The manure and urine deposits are not concentrated and are processed by dung beetles. There are no manure piles or lagoons that release methane, as occurs in CAFOs[26]".

One of the farms known nationally and internationally for its regenerative land management system is White Oak Pastures (WOP) in the US state of Georgia. The farm produces grass-fed beef, lamb and pork as well as pastured chicken and poultry. The environmental sustainability consultancy Quantis conducted a Life Cycle Analysis (LCA) of the farm's products which "evaluates the total environmental impact of a product over its entire production (and/or consumption) chain, allowing for a comprehensive comparison of alternative ways of meeting human needs and economic functions"[27]. The main finding of the peer-reviewed study relating to beef: "The net result is that WOP beef has a carbon footprint 111% lower than a conventional US beef system. The WOP system effectively captures soil carbon, offsetting a majority of the emissions related to beef production". And the authors conclude: "Within our margin of error, there is a potential that the WOP beef production is climate positive. This would be very rare and it is unusual that there is more benefit to producing something than to simply not produce".

Grazing isn't just for beef cattle, there is no reason why dairy cows shouldn't be out on pastures, too. And they used to be. In 2023, an analysis by Greenpeace Germany showed that while in 2010 42% of dairy cows in Germany had access to pastures at least part of the year, in 2020 that was only true for 31% of cows[28]. The reason: very high yielding dairy cows need to be fed concentrate because the day isn't long enough for them to eat and digest enough grass to provide them with the amount of calories they need.

But what if the milk yield was lower, would it be possible to reduce the amount of concentrate feed and have dairy cows out on grass instead? A three-year study in Germany tried to find out. I summarised the results for the Sustainable Food Trust in 2022: "The

122 participating dairy farms fell into one of four groups: certified organic farms feeding either low or high amounts of concentrate feed and conventional farms with either low or high use of concentrate feed. The average herd size was 49 cows, the average milk yield per cow per year was a little over 6,000 kg.

The big surprise from the study was the economic data: dairy farms using little concentrate feed earned less from each cow, but their costs were so much lower that they had a healthier bottom line – over all they proved to be more economical. (...) While organic low concentrate farms mostly saved labour costs, conventional low concentrate farms spent almost 40% less on pesticides and 20% less on fertiliser. And, of course, there are measurable benefits for the environment when chemical inputs are substantially reduced.

The study also shows that not all cows do well on a reduced concentrate feed diet. Of the successful farms, where animals did well on little or no concentrate feed, 80% kept regionally adapted dairy breeds, crossbreeds and dual purpose breeds such as Simmental. "For a low concentrate feed regime, it's important that the dairy cows are long-lived, healthy and robust,' writes Karin Jürgens (Unabhängige Bauernstimme, October 2021), one of the authors of the study. The cows need to be able to graze well and be selective in order to make optimum use of the available feed, she says. 'That's why these farms use cross-breeds adapted to the system or animals bred on their farm.' While Holsteins are the most common dairy breed, followed by Jerseys, the scientists found a high diversity of dairy breeds on the low concentrate feed farms, with 15 of the 21 dairy breeds registered in Germany represented. Just 15% of the low concentrate farms in the study kept only rare breeds.

The success of the low concentrate system depends on how much milk the cows can produce from grass – which includes grazing permanent pastures, clover and lucerne (alfalfa) as well as feedings of hay. On conventional farms 72% of the milk came from grass, on organic farms it was 81%. (...) These farmers do valuable work for the environment, says Jürgens, but even though it is a profitable way to produce dairy, the overall farm income lies below the national average. One reason is that the number of cows will depend on the grassland available, which of course cannot be scaled up. Jürgens' conclusion: the work these farmers do is not sufficiently supported"[29].

Environmentalists and a number of farming organisations are lobbying decision makers in Brussels to increase CAP payments for providing ecoservices through good grazing management. There is also an ongoing discussion about the introduction of premium products with a 'pasture milk' label which could compensate farmers somewhat for the lower milk yields in a grass based dairy system.

And this gets to the heart of the problem of GHG emissions, not just from ruminants. "Livestock becomes a menace only when, as now, we feed half the world's cereal and most of the world's soya to animals that should be getting the bulk of their nutrients from grass and leftovers. We burn vast quantities of oil to grow that cereal and fell forests to grow the soya – and so contribute twice over, and prodigiously, to global warming. Yet the cereal and soya we feed to the animals need not be grown at all – or if it is, then we could be eating it ourselves. Again, we find it is not the animals that are at fault but the system"[30], writes Colin Tudge.

At fault is an industrial agricultural system with farms that raise animals in confinement. According to Joyce D'Silva, 98% of global soybean meal is used as animal feed. "Corn (maize) is the most widely produced feed grain in the United States, with around 36% of the crop providing the main energy ingredient in livestock feed[31]" In her calculation, 36% of the world's crop calories are fed to animals. In a conventional system, production of crops takes huge amounts of chemical fertiliser and pesticides, both fossil fuel based products. More fossil fuels are needed to run the machinery needed to spread them on the crops.

Much of the blame cows get is simply a case of false accounting: "This ranking in which cattle and beef always fare worst, has two main flaws: firstly, there is no differentiation between energy intensive, resource draining agricultural systems and sustainable ones. Secondly, the relevant consequences and costs of industrial animal production are externalised[32]", says Anita Idel. "Growth in demand is what turns a sustainable system into an unsustainable system[33]", says Tara Garnett, the founder of the Food Climate Research Network. "We show that tackling climate changing emissions will require us to reduce the number of animals we rear and to change the way we rear them; and we will need to eat fewer foods of animal origin. This said, livestock production can also very much form part of the solution – farm animals can help create a resilient, sustainable, biodiverse food system, if we rear them in the right way and at a moderate scales[34]".

Colin Tudge puts it somewhat more pointedly: "Now we come across a huge serendipity – or rather, two huge serendipities. To be sure, if we fed livestock only on grass or leftovers and (genuine) surpluses, then we would not produce as much meat as we do now. We might indeed produce far less, at least in some places. People in high places, keen to maximize profits by maximizing output, are wont to tell us therefore that if we did not grow many millions of hectares of soya and maize for cattle, pigs, and poultry, our diets would be intolerable – all lentils and garlic bakes. Again our rulers display their ignorance. They clearly know nothing about food. For the kind of farming that really would feed us all focuses on arable and horticulture with livestock in supporting roles and so produces 'plenty of plants, not much meat and maximum variety': and these nine words 'plenty of plants, not much meat and maximum variety' – summarize all the most worthwhile nutritional theory of the past three decades"[35].

Land management team here to help.

Integrating livestock into arable systems and livestock as part of a well-managed grazing system benefits soil health and, on permanent grassland, enhances carbon sequestration. Aren't such ecoservices enough? Do we need to slaughter animals and eat them?

Questions such as these were discussed in January of 2024 at the 'Livestock in the Landscape' session at the Oxford Real Farming Conference, ORFC. Livestock are not only needed on agricultural land, they are essential for habitat conservation, says Lindsay Whistance, senior livestock researcher at the Organic Research Centre, ORC. Cattle are such an important 'land management tool' that DEFRA (Department for Environment, Food and Rural Affairs) dedicates a website to habitat improvement through livestock. "The best way to conserve some habitats is by traditional grazing. This is also known as conservation grazing. Habitats suitable for conservation grazing include: grassland, heathland, wood pasture, coastal and floodplain grazing marsh, including areas with breeding and wintering wetland birds, fen, scrub and scrub mosaics, saltmarsh and sand dunes.

They can be home to a large number and range of species, including threatened or rare plants, invertebrates and birds.

**Hardy and happy
outside year round.**

Conservation grazing creates vegetation at different heights, and small areas of bare ground. This makes it suitable for a wide range of wildlife in different habitats. It allows wildflowers to grow, flower and set seed each year. This provides pollen and nectar for invertebrates and increases invertebrate food available for birds[36]".

The website also provides references and further information from choosing livestock to grant options under SFI (Sustainable Farming Initiative). And then there is 'manage your stocking levels': "Your habitat may not support the nutritional needs of livestock all year. You'll need to remove livestock at these times or give habitats a rest period. You'll need to identify other grazing land to move livestock to at these times".

In an environmentally balanced ecosystem no species will grow exponentially, says Whistance. Population numbers are kept in check by predators, through disease – in the UK, rabbit numbers have been decimated through the spread of myxomatosis – or otherwise, like the UK deer population, is at risk of starvation[37]. "We have to recognise that population control is a fundamental requirement for most species on the planet. Apart from apex predators (such as wolves, lions or tigers) there is no evidence that any species can control its population according to availability of food sources. We are all reliant on some external control or self-imposed control", says Whistance. Which means: if we were not to control cattle numbers through slaughter or reintroduce a predator species like wolves to the UK, cattle would succumb to disease or die of starvation. If done humanely, slaughter is by far the kindest, quickest option.

Vegans may nevertheless object to the consumption and use of animal derived products. In regard to meat and dairy products from animals raised in confinement I couldn't agree more. I, too, don't want to eat meat from animals that barely had room to move and had a long, stressful journey to an abattoir. But, as Lindsay Whistance put it: "There is a place for human food, and animal derived products, but this should come as consequence of planetary health and ecosystem health first and foremost[38]".

Of course I respect anyone's personal dietary choice. What I find problematic however is the 'pick and choose' attitude in the definition of what qualifies as 'vegan'. On the Vegan society's website it says: "Veganism is a philosophy and way of living which seeks to exclude—as far as is possible and practicable—all forms of exploitation of, and cruelty to, animals for food, clothing or any other purpose"[39]. 'As far as possible' is a neat copout from having to admit that the production of grains, legumes, fruit and vegetables, too, is impossible without animals, many of which die in the process.

In a confrontation with a vegan protester, the rancher John Dutton, the main protagonist of the US television series 'Yellowstone', made the point rather succinctly: "To plant the quinoa, the sorghum or whatever the hell it is you eat, you kill everything on the ground and under it. You kill every snake, every frog, every mouse, mole, vole, worm, quail, you kill them all. So, I guess the only real question is how cute does an animals have to be before you care if it dies to feed you"[40].

Vegans don't eat honey, but they will consume almonds (and almond milk), 80% of which will have been grown in California. Scientists at the University of California, Davis calculated the numbers: "In late February, about 90% of all honey bees in the United States are in California, pollinating almonds. They come from all over the country. (...) Almond growers typically use two bee colonies per acre. Each colony contains about 20,000 bees, so there are about 48 billion bees pollinating California almonds right now"[41]. According to the USDA (US Department of Agriculture), a third of all commercial colonies, die every year[42]. Colony Collapse Disorder is the catch-all term for stress through transport, exposure to pesticides and disease.

Greenhouse tomatoes can also only be grown by exploiting animals: without bumblebees pollinating the plants there would be no fruit. Growers annually order bumblebees through the mail – the precise number needed for their greenhouse space. Such colonies are raised in factories and mostly end up in environments where they are not native. Bumblebees escaping from a greenhouse environment can spread disease among native species.

These are just two obvious examples to illustrate John Dutton's question: how cute does an animal have to be?

According to the British Vegan Society, vegans also shun clothing made from animal derived materials. Wearing cotton and linen in a hot climate is sensible, in a cold climate, warmer materials are needed. The only alternatives to wool are synthetic fibres, all of which are made from fossil fuel. When they are washed in a washing machine they shed microfibres, the type of microplastics that are polluting soils, waterways and oceans. They can be found in fish, shell fish and other sea creatures and in our food. "Researchers have found microplastics damage human cells, decrease reproductive health, and disrupt the endocrine system. Microplastics also act as a vessel for harmful substances to enter the body as they can absorb chemicals linked to cancer and weakened immune system,"[43] according to the non-profit organisation Earth Day.

Lastly, a quick word about the arguments of the author and vegan campaigner George Monbiot. In his book 'Saying No To A Farm-Free Future'[44], author Chris Smaje takes on Monbiot's claims that we could swap agricultural produce for lab-grown calories by simply doing the math: "The RebootFood campaign that he (Monbiot) is fronting invites readers to imagine rewilding three-quarters of the world's farmland and producing the entire world's protein on an area the size of Greater London".

For his film Cow Apocalypse Monbiot visited a company that is pioneering the cultivation of a bacterium, *Cupriavidus necator*, in bioreactors. Fed with hydrogen and oxygen, the bacteria multiply and produce a protein rich soup which supposedly can be made into something resembling food. Smaje dives deep into Monbiot's calculations and concludes that on a global scale this type of 'food' production would require around 43% of the world's electricity consumption. If we want to save the planet, we can no longer rely on fossil fuels, but need to either build more nuclear power plants or use solar and wind energy. A bioreactor capable of churning out 43,000 tons of microbial protein would need an area of 3,000 hectares covered with PV (photovoltaic) panels to produce the necessary energy. "If you aggregate that up to cover the protein needs of the entire US population, there would need to be about 140 such facilities, with a total annual capital cost of $12.9 billion and over 400,000 hectares of PV panels". Scale this up to the protein needs of 9 billion people, and there isn't much land left to re-wild. And as Smaje rightly points out: this is simply the energy requirement for operating the bioreactors. It does not take into account the energy needed for the production of the steel vats which last for a maximum of 25 years, nor for the millions of PV panels or the environmental and energy costs for necessary mineral production through mining.

At the 2024 ORFC Michael Lee was one of the participants in a session titled: What role for grazing livestock in a warming world? Lee is a professor at Harper Adams University, an expert in sustainable livestock systems and livestock methane emissions. At the end of the session he was asked why the huge climate potential of grasslands were so often ignored and cattle vilified. This is what he replied: "I, too, have asked this question and I got to a view that there is unfortunately a group who are ideologically opposed to livestock. (...) What we continually see is the continual use of global averages for beef. We continually see the use of carbon dioxide equivalents per kilogram of product even though we know that that's not appropriate in terms of the balance and we continually see gross evaluation of ruminant systems instead of net emissions, i.e. taking into account soil and pasture growth etc. I think the only logical outcome of this is that ideologically the narrative is that livestock are wrong for a sustainable planet and therefore it is an inconvenience in that narrative to look at improvements. It is an inconvenience that UK beef systems have a lower carbon footprint than the global average – so we use the global average. It's an inconvenience that carbon drives soil health and carbon stocks are driving that improvement. It's an inconvenience that ruminants are the heart of rural communities. It's an inconvenience that animal sourced food and in particular red meat are the best provider of heme iron and we have 55% of young women in the UK who are iron deficient. It's an inconvenience and therefore is ignored. Because it goes against the narrative which is an ideological viewpoint for the removal of livestock for the planet[45]".

Chapter 2

Making the case for small abattoirs in the UK

"We live in interesting times, full of uncertainty, and with the future direction of UK food and farming at the centre of many of these unknowns. It feels clear to me now, more than ever, that we need to expand our options and broaden our markets, local and global, by producing food with great credentials." This is the opening sentence of the foreword Phil Stocker, the chief executive of the National Sheep Association wrote for the Sustainable Food Trust's 2018 small abattoir report 'A Good Life and a Good Death: Relocalising Farm Animal Slaughter'[46]. "The growth of the local food market and interest in 'single farm provenance' will be an increasingly important part of what promises to be an exciting future for the UK livestock sector. In particular, it could provide a valuable lifeline for our hard-pressed upland livestock farmers producing some of the best grass-fed beef, lamb and mutton anywhere in the world".

As he wrote the introduction in 2018, two years before the UK left the European Union, and two years before the pandemic struck, Stocker probably had no idea how prophetic his words would be. Shoppers, supermarkets and suppliers – nobody was prepared for the British government sending everyone into lockdown in the third week of March 2020. And while supermarket shelves remained bare, sometimes for weeks, farmers stepped into the breach and delivered food: through box schemes, for collection at designated pick-up points, in farm shops, and through vending machines. It took just a few weeks for a functioning local food system to emerge. Farmers kept people fed by coordinating with other farmers. In communities up and down the country, volunteers helped aggregate orders and made sure they could be picked up or delivered. Farmers blogged, tweeted, shared recipes for those cooking from scratch for the first time, and they kept spirits up with live feeds from lambing sheds and birthing barns. In rural areas, consumers connected or reconnected with farmers in their neighbourhood, while city folk, often for the first time, started to actively think about the fact that the food on supermarket shelves was actually produced on a farm.

Since 2022, supply chain issues, import problems due to Brexit and, of course, rising energy costs and inflation continue to plague many industries, not just food producers and supermarkets. Mostly, shelves are reasonably well stocked again. But even though more shoppers continue to buy some produce directly from farms, local food systems no longer make great strides. "The ambition of the local food sector is stymied by systemic barriers" found a study by the University of Sheffield[47] published in May, 2022.

The barriers for meat producers are particularly high. "The infrastructure required to deliver local traceable meat consists of specialist smaller local abattoirs, which enable farmers to slaughter their animals relatively close to the point of production, stipulate how they want things done, and be sure about getting their own products back. Yet these small independent abattoirs, proportionate to the scale of farms in their vicinity, are in the process of disappearing without a trace," wrote Stocker in 2018. 'A Good Life and a Good Death' tracked the numbers: "In the 1930s there were said to be around 30,000 red meat abattoirs in the UK. By 1971 this had dwindled to 1,890, and by 2003 there were just 320. By July, this had further eroded to 251[48]."

For her thesis, Amy Quirk, a University of Greenwich agricultural graduate, took a look at small abattoirs licensed to slaughter beef cattle and the impact abattoir closures have on rare and native breed horned cattle in England. According to FSA figures from December 2023, there were 122 beef abattoirs open in England, down from 165 in 2019 and 289 in 2000.

Of these 122 only 115 are actually operating, Quirk told me. "Seven hold 'dormant licenses' which means that should they decide to reopen within three years they could re-instigate the old license following checks[49]". Quirk is especially interested in rare and/or native breeds. Many abattoirs are either not willing to take such animals or not suitably equipped. The BSE crisis ended years ago, but it still takes a special permission to slaughter animals older than 30 months because the spinal cord has to be entirely removed. "In conservation grazing animals will often be older that 30 months. Smaller abattoirs that do take horned cattle often have restrictions put in place because of health and safety concerns or the built of the stun box. So, if you keep the rare, native horned breeds your options for slaughter are extremely limited and vital for profitability as a speciality product". Quirk found that out of 115 abattoirs in England only 22 take horned cattle over 30 months old. She has put together an interactive map[50], a useful tool for farmers to check what type of abattoir is within a certain radius of their farm.

Not only abattoirs are closing in record numbers. The same goes for independent butchers. While there were still 15,000 butchers in 1990, their number fell to just 6,000 in 2015. And according to figures by the Agriculture and Horticulture Development Board from 2019, the number of independent butchers has gone down 60% in 25 years with more than 100 shops closing per year[51].

One of these was J W Mettrick & Son Ltd in Glossop in Derbyshire. John Mettrick is a fourth-generation butcher and slaughterman. The butcher shop on the high street used to sell only meat from animals that Mettrick and his team had slaughtered. In addition, the family has long provided custom butcher service (also known as 'private kill') to local farmers: they could deliver individual or a small number of animals to the abattoir and would be certain to receive the meat from their own animals back – as a whole carcass or as sides of beef, or butchered and packed, ready for sales off farm or online.

For farmers, this service is absolutely crucial if they want to add value through direct sales. And at no time has this been more relevant. Since the UK left the European Union, farmers receive decreasing amounts of money under the Basic Payment Scheme (BPS). Agriculture is a devolved policy area and for farmers in England the government will continue payments at a reduced rate with the intention to end them completely in 2027. BPS payments will be replaced by the Environmental Land Management scheme (ELMs). The terms for the different programmes are a work in progress. One of the programmes, the Sustainable Farming Initiative (SFI) has been open for farmers to join since 2023 and is still evolving. A number of practices eligible for payment have been added and some measures are now paid better, others earn less money. One thing is clear nevertheless: ELMs can make up for some, but definitely not all BPS payments that farmers received, and that will leave hill and upland farms in particular with a huge gap in their finances. Diversification is one of the ways to make up for the shortfall, and for small and mid-sized mixed farms, direct sales of meat could be an excellent source of income. But farmers can only realise this potential if they can find an abattoir that will slaughter and butcher their animals.

The financial benefit for farms and farmers is just one aspect of livestock rearing and meat production. As important is animal welfare: transport is stressful for animals. Some may be accustomed to being transported on a trailer over short distances from one pasture to another, but long journeys cause stress. And with more and more small abattoirs closing, distances increase considerably – animals now often have to travel for several hours. Or, as Quirk told me, farmers raising Longhorns have been given the option by abattoirs to de-horn their cattle instead of driving them across the country to an abattoir that takes horned animals. It's a terrible choice to have to make because de-horning even under local anaesthesia is traumatic for the animal. Conditions on a truck are often cramped and usually animals are not fed or watered before they go to slaughter. Sheep and cattle are herd animals and being taken out of their familiar group will make them anxious.

Long journey times are of particular concern for organic livestock farmers. John Pawsey, a farmer in Suffolk, told me that for a time he sold the meat from his organically raised sheep and lamb as conventional because the nearest certified organic abattoir was several hundred miles away and he wouldn't have any of his animals travel that far.

The closure of a small abattoir has many indirect consequences for rural communities, the environment and the food system. If farmers can't direct market their meat or sell it through farm shops and local butchers, in the long run, this will have an impact on the rural economy: less money will be spent in local shops and jobs will be lost. Customers will have to travel further to do their shopping, which increases greenhouse gas emissions and air pollution. Not only will they find it harder to purchase local, farm-traceable meat, the quality of the meat may also be poorer – the less stress an animal experiences before and during slaughter, the better the quality of meat.

With this in mind, the All-Party Parliamentary Group for Animal Welfare (APGAW) recommended in their 2020 report[52] the formation of an Abattoir Sector Group (ASG) with the mission "to support and develop a thriving network of local abattoirs across the UK, which is essential to maintain high animal welfare standards, support sustainable farming and meet the rising demand for local meat"[53]. Among the driving forces in the ASG are the Sustainable Food Trust, Fir Farm, the Rare Breeds Survival Trust, the Prince's Countryside Fund, National Craft Butchers, Animal Health & Welfare Board England. The group is chaired by John Mettrick, the butcher and abattoir owner from Glossop.

Because of his tireless fight for the survival of small abattoirs, Mettrick made national headlines when he had to close his abattoir on 31st August, 2022. On the morning of that day the BBC's Farming Today programme aired a conversation between John Mettrick and presenter Anna Hill. The interview once again made clear what's at stake and how emotional the issue is for all concerned.

Anna Hill: "Later today, a Peak District butcher and small abattoir owner who led a nation-wide campaign to save businesses like his, will be closing his abattoir for the last time. John Mettrick, whose family have run the operation in Glossop for more than 100 years (...) According to the Food Standards Agency, small abattoirs are closing at a rate of 10% each year. Mr Mettrick believes that part of the problem is that regulations have been designed for large operators and are difficult for small abattoirs to manage. He told me how his family are feeling about the closure."

John Mettrick: "Very upset, to be honest. You know, father is crying, he is 83, sister is crying, it is gutting, really, and particularly gutting for all those farmers whom we are having to turn away, knowing the effect that might possibly have on their businesses where they might actually have to stop trading because the distances to an abattoir that will provide the service we do could be too great for them."

AH: "Let's have a look at the causes of this. You have been reported as saying it's down to two things: one is inappropriate and burdensome regulation and inexperienced on-site official veterinarians. They are the government vets. So, let's take that first of all. Why is the regulation something that forced you to close?"

JM: "During COVID, we all know there was a huge demand for local food. And we actually increased the throughput of our abattoir to actually cope with that demand from the public. And what's happened is the Food Standards Agency have counted those two COVID years and have come to an assessment that our official veterinary presence in our abattoir should be increased from what was a part-time presence, where the vet would look at the animals live and then come back the next day to inspect the carcasses, to full-time veterinary supervision. The main slaughterman, by the way, who won the Abattoir Apprentice of the Year, walked out and two butchers followed as well."

AH: "You are describing your own situation and that's very specific to you. Do you think the same thing is happening in other small abattoirs as well?"

JM: "I know it's been happening in small abattoirs for years. The OV [oversight veterinarian] presence is very important in abattoirs, I don't want to say they are not important. But the level of OV presence in small compliant plants is excessive. We desperately need new legislation. We're not asking for handouts, we are asking for risk based, proportional regulation so businesses like ours can go on into the future. Because if you talk to a young person and try to tell them that they are going to be on CCTV camera all the time, every stun of an animal they make is going to be recorded and in the cloud, and then they have an inexperienced vet standing in front of them timing them on a mobile phone – who wants to work in that kind of an environment?"

AH: "You've been one of the main campaigners for small abattoirs, haven't you? And yet you are having to close your doors!"

JM: "I've really tried hard, Anna, I've given them examples, I know that people in the FSA are upset by this situation, and they do care, but until this legislation comes in to change it, the Food Standards Agency aren't going to do anything about it."

AH: "But it's not just the business, is it, it's the family."

JM: "Yeah, it is. That's the worst part. You know, when you have to talk to a slaughterman who has done the job for 21 years, you are telling him why you are having to shut and he says 'Where can I go, what can I do?' And you know, my dad and my granddad and his father before him, my great granddad, we are all butcher slaughtermen. So, it is a big departure for us. We've tried everything that we can and I will still stay on and fight the corner for the small abattoirs even though I've lost my own."

AH: "We've asked for a response from DEFRA. In a statement it said: 'We are working with both the Food Standards Agency and the Rural Payments Agency to streamline administrative burdens and our small abattoir working group is engaging with industry to ensure we are taking a strategic view of the issues the sector is facing.'"[54]

In September 2022, the Sustainable Food Trust and the National Craft Butchers put together a survey for abattoir users. The results show that since the publication of the 2018 small abattoir report 'A Good Life and a Good Death', the situation hasn't improved. "According to Food Standards Agency figures, small abattoirs are closing at a rate of 10% each year – meaning that within a decade they may disappear altogether"[55], the report finds.

During the SFT Local Abattoirs session at the 2024 Organic Real Farming Conference, ORFC, farmers, butchers and members of the Abattoir Sector Group took stock.

In her summary of the session, SFT's Megan Perry quotes Phil Scott from the wholesale meat supplier Lake District Farmers for fell farmers in Cumbria. "Despite there being 5,242, mainly livestock-based, farm holdings in Cumbria with a 12,500 strong workforce, there are only two abattoirs offering private kill, both in the south of the county. Following the closure of Black Brow abattoir in Wigton last year, farmers are now travelling from as far as Scotland down to Airey's abattoir near Grange-over-Sands to take their animals to slaughter, but this is having a knock-on effect for local customers, with a local butcher's shop unable to get their animals slaughtered in the run up to Christmas. He emphasised the need for more skilled staff, both slaughterhouse workers and butchers, to ensure the sector thrives, and called for more apprenticeships and better advertising of the sector as being a good career path for young people"[56].

To give the sector a boost, DEFRA launched the £4 million Smaller Abattoir Fund at the end of 2023. John Powell, the Head of Agricultural Teams at DEFRA provided an update at the ORFC. Eligible abattoirs can "apply for grants to improve animal health and welfare, increase productivity, add value to products or contribute to new technology and innovation"[57]. Grant money cannot be used to supplement wages or cover rent and running costs such as electricity.

At DEFRA, there is also work being done on a rule change that could help small abattoirs significantly, says Perry. "John Powell also shared that Government have begun pulling together a case for adoption of the '5% rule', and that Ministers are keen to see this happen quickly. The rule is a flexibility that exists in EU legislation and enables smaller abattoirs to slaughter up to 5% of the total national throughput without triggering full veterinarian presence and charges. Failure to adopt this rule has stifled productivity and growth in the sector as some of the smallest abattoirs opt to keep their throughput under the 1,000 livestock units per annum allowance in order to keep inspection and charges to a minimum".

For some abattoirs help may come too late. In January, Long Compton Abattoir (LCA) which slaughters animals for farmers mainly in Warwickshire, Oxfordshire and Gloucestershire has started a campaign to raise money. If no investors come forward the Long Compton will close.

In February 2024, McIntyre Meats in Yorkshire announced its closure. On the BBC's Farming Today, farmer Graham Bottley explained what the closure means for him and his business. He raises Swaledale sheep and provides mutton to private customers and high-end restaurants. He takes between one and three animals for slaughter at a time, up to now a journey of 10 to 15 minutes. After the closure the two 'nearest' abattoirs will be a 60 to 75 minute drive away. "It's not really feasible in terms of time, the economies and it's not good for animal welfare either[58]", Bottley told Farming Today. Given the circumstances, he may have to stop producing mutton. "The supply chain issues for me are that if I don't have ready access to an abattoir, I don't think I can continue with the business in this form. (…But:) This isn't really just about me, this isn't really just about McIntyre's down the road, this really is a countrywide issue. Every abattoir that closes rips the heart out of all of those small businesses that rely on the abattoir. So, as these small businesses close the remaining butchers close. And that reduces choice and so on and so forth. And then eventually you end up with just a centralisation of production which I don't think is good for anyone. And it's certainly not good for farming".

The situation in Britain is not unique. Across western Europe and the US small abattoirs have been closing in record numbers and farmers face similar problems. But unlike their British colleagues they seem to be able to find solutions – individually for their farm, co-operating with other farmers, working with the industry and/or successfully lobbying for change in government regulations. Circumstances vary, no single solution fits all.

However, all approaches have one factor in common: their success hinges on the availability of well trained, artisan butchers, and their numbers have been in decline, too. But with a growing focus on locally sourced food, animal welfare standards and humane slaughter, more young people are beginning to see butchery once again as a career option.

Chapter 3

Humane, local and profitable – how it can be done

The first chapter set out why we need animals in agriculture, the potential of grazed grassland for the environment and the role of animals on farms with arable land. Chapter two gave an overview of the situation for small abattoirs in Britain and the implications their shrinking number has for farmers and animal welfare. Whether the animals are raised for meat or dairy, whether their main role is habitat preservation or whether it's a combination of these functions, at some point the animals will have to be slaughtered[59]. Their meat can provide us with (a lot) of nutrient dense, healthy food as well as other products such as wool and leather. "If people did not obtain these good from livestock, they would need to be produced by some other means and this would almost inevitably incur an environmental cost"[60], says Tara Garnett, the founder of the Food Climate Research Network.

In the farming systems described, animal welfare is paramount. The animals have lived 'a good life', not in confinement, but grazing or, in case of pigs, rooting for food and able to wallow in mud. Such life should be followed by 'a good death'. Animal welfare does not end at the farm gate when the animals are loaded onto a trailer that takes them to an abattoir. The end point is a humane slaughter process.

Farmers employing sustainable and environmentally friendly farm practices, work with nature and put animal welfare at the top of their agenda, nevertheless run a business and have to make a living. Or as Cynthia Daley from the Center for Regenerative Agriculture at the California State University in Chico put it: "If you are not economically sustainable you are not sustainable at all".

On a diverse, mixed farm or in a purely grass based system profitability cannot be achieved through ever increasing yields, there are clear limits to how much meat and/or milk can be produced. Tara Garnett suggests to raise 'livestock on leftovers'. "Such an approach seeks to work with what animals are good at – making use of marginal land and feeding on by-products that we cannot eat. It considers what the land and available by-products can sustainably support – and then assesses, on that basis, how much is available for us to eat. It takes land and the biodiversity that it supports as its ultimate constraint. (…) With this approach livestock can be integrate into a landscape so that they help store carbon in the soil, enhance the biodiversity of local ecosystems and make use of the leftovers from other food and agricultural processes. There will be a need to focus research on breeding programmes that emphasize robustness and flexibility and that improves the ability of ruminants to survive on marginal lands. (…) We need also to consider how livestock can be integrated into arable farming systems in the developed world as part of a mixed livestock – crop rotations and livestock – agroforestry systems. (…) This 'livestock on leftovers' approach is likely to provide us with far lower quantities of meat and dairy than that afforded by intensification – but the system actively helps deal with the problem of climate change, while the efficiency approach is simply geared towards minimizing the damage that livestock cause"[61].

John Webster, professor emeritus at the University of Bristol and founder of the UK Farm Animal Welfare Council, also considers grazing cattle on well managed pastures and in silvo-pastoral systems to be highly sustainable and environmentally friendly, but not sufficiently profitable: "Their output, measured simply in terms of food sales, is likely to be too low to meet the need of the farmers to sustain a reasonable standard of living"[62]. That certainly is true in the world of supermarkets and cheap meat. We, as a society, cannot expect farmers and ranchers to deliver numerous environmental benefits without being able to make a profit. Direct marketing can provide good opportunities – but to make use of them means having access to humane slaughter facilities and butchers.

It is beyond the scope of this book to reflect on broader ethical and historical questions as they are discussed for example by Kristian Bjørkdahl and Karen V. Lykke in Live, Die, Buy, Eat: A Cultural History of Animals and Meat or Noélie Vialles' Animal to Edible.

The topic here is how the 'good life' of an animal can end in a humane way, with animal welfare being paramount until the very end. John Webster sees the Five Freedoms as "an exceptionally valuable tool for assessing the welfare state of an animal, or a herd of animals, at a moment in time".[63] The Five Freedoms are to be (1) free of hunger and thirst, (2) discomfort, (3) pain, injury and disease, (4) fear and stress and (5) able to exhibit natural behaviour. Webster argues: "No system of animal production can guarantee absolute freedom from suffering and ill health. However, many risks to health and welfare arise as a *direct consequence of the management system*"[64].

Webster assesses different systems; for beef cows and their calves he concludes: "It is my belief that beef cows who spend most of the year on well-managed pastures in temperate climates, give birth to one calf per year and feed it for about six months may experience a better quality of life than any other class of animal reared under commercial conditions. (...) Suffering occurs when an animal is unable to cope (or has great difficulty in coping) with stress. Beef cows on well-managed pastures all over the world are well able to cope with natural stresses most of the time. This reflects not only the extent of their physiological powers of adaptation. They are also able to enjoy perhaps the most important of the freedoms: freedom of choice[65]".

'Animal welfare' is commonly used in the context of an animal's life, but several of the Five Freedoms can be applied to the *end* of an animal's life too: the animal should not experience hunger or thirst, no fear or pain, and as little stress as possible. By killing an animal on the farm or shooting it in the herd all stress can be avoided. The next best option is slaughter in a small abattoir with well-trained slaughtermen and butchers, in as close a proximity to the farm as possible. Depending on circumstances, some animals may be used to travelling in a trailer and will therefore cope with transport better than others, but loading, unloading and the discomfort of being in a moving vehicle are stress factors.

How such humane slaughter systems can work, what conditions need to be in place, the legal framework needed and the question how farmers and butchers can stay profitable nevertheless are the central topics of this book.

Across the world, farmers, ranchers and those who buy animals off of them have to figure out where and how to get these animals to slaughter. In many countries, the number of small abattoirs is decreasing and solutions have to be found.

I have chosen to focus on two specific aspects: how farmers and butchers in Germany are tackling the problem and, in the US, on the developing niche meat sector.

There are several reasons for this choice: an EU regulation introduced in 2021 provides a legal framework for German and other EU farmers to slaughter animals on farm not just for home consumption, but to sell the meat into the food chain. Introducing similar solutions in the UK is a distinct possibility, even after Brexit.

The different approaches described here could be altered, amended or simply serve as a starting point to come up with solutions in the UK.

Humane slaughter is a very important issue, but without market access, meat production and processing cannot be sustainable. In the Pacific Northwest of the US, a niche meat market is developing. Farmers, restaurants, chefs and retail outlets are involved in building local and/or regional food chains for meat.

Part Two starts off with a short overview of the protocol for on-farm slaughter that has been formulated on the basis of the amended EU regulation. What are the obligations of farmers and vets and how does it work in practice.

Animals can now be shot in the herd if they are kept on grass year-round. On a Demeter certified dairy farm near Lake Constance cows spend the coldest and wettest weeks in the barn and the farmer has learnt how to stun an animal before slaughter to keep even the last few seconds of its life stress free.

But on-farm slaughter can be really expensive. A butcher in North Rhine-Westphalia therefore offers a bespoke service which reduces the costs for farmers and increases his profitability.

Where and when are small abattoirs the solution? Chapter 6 presents three case studies. One small abattoir is owned and run by a group of beef and dairy farmers. The second facility was newly built because a butcher with a flourishing butcher shop decided to expand his business by adding custom slaughter services. The third case study describes a pig farmer who made humane slaughter an absolute priority and invested in a custom built slaughter facility on her farm.

Even the most modern abattoir can only be as good as the staff operating it. Chapter 7 explains how the German formal three-year butcher apprenticeship programme works. At a regional college, third year students talk about their motivation and aspirations. Apprenticeship programmes include both, class based learning and practical training on the job.

In Bavaria, a construction company has invested in a unique venture: a farm business with a custom built small abattoir, restaurant and shop offering apprenticeships for butchers, agricultural workers, clerical and catering staff.

As with all farms and businesses I visited, meat quality is of great importance. The Chapter 7 explores the connection between humane, stress free slaughter and meat quality as well as the nutritional value in particular of grass-fed beef.

Part Three opens with a trip to one of the few remaining prairies of the western United States, exploring the importance of native grasslands and casting a short look back at the Dust Bowl era. Today, beef cattle spend little or no time grazing. Instead they are 'finished' on a diet of maize (corn) and other grains in feedlots. Four big meat processors dominate the market, operating industrial scale slaughter facilities. Farmers and ranchers who don't see their animals as raw material to be fed into this system or who want to market meat directly and/or locally have to get creative: from working with a mobile slaughter unit to building and running a small abattoir.

In the Pacific Northwest, close to urban centres such as Portland and Seattle, local, sustainable niche meat supply chains are emerging. Six examples show how farmers, butchers and chefs work together: adding value to meat, creating 'consumer ready' meat products, and finding ways to allow better market access. With the support of NMPAN, the Niche Meat Processor Assistance Network, a University of Oregon Extension based community, a number of small processors are facilitating such efforts.

Like the UK, the US have no formalised system to train butchers, most training happens 'on the job', someone is a butcher-slaughterman because they do the work. Four examples demonstrate how virtual and practical learning can be combined and why management and marketing need to be added to the skills list, too. Processors organised in the North West Meat Processors Association, NWMPA, have gone a step further and are now offering a three-year paid apprenticeship that not only focuses on butcher skills, food safety and food processing, but also on managerial skills which will enable graduates to set up their own business. It is an effort to emphasise the value and skill of craft butchers while making the niche meat industry more efficient and profitable.

Livestock holders in the Pacific Northwest may have better access to processing facilities, but marketing can still be a challenge. In the Midwest, a farmer has embarked on building a supply chain to help producers of grass-fed meat to get it to market.

Part Four looks at which of the models, concepts and ideas described in the case studies might be suitable for the UK. But in addition, we need a change in perception: cattle in particular need to be seen as 'land managers' essential for climate mitigation, while the availability of humanely slaughtered meat is essential – for farmers, local communities and consumers. And if we are to get to net zero, we need to utilise the carbon sequestration potential of grassland by having cattle graze it. Livestock farmers are providing a public good for which they should receive public money. Meat production is a welcome by-product.

What you will not find in this book are statistical analysis, graphs and diagrams because this is a new and emerging field. The EU regulation that allows for slaughtering animals on a pasture or on farm was only passed in 2021. The number of MSUs operating in any country is limited. Small abattoirs owned and run by farmer cooperatives are a rarity. The numbers are far too small for any valid statistical analysis or meaningful comparison.

This book is the result of numerous trips, within Britain, in Germany and in the US. On all these trips I travelled with Martin Kunz, my husband and the man who takes the pictures, some of which are included here, many more are available on my website[66].

As you may already have noticed, my writing is something of a 'linguistic crossbreed'. I use both British and American source material and while one talks about fertiliser and maize, the other spreads fertilizer on corn. A quote is a quote and that includes the spelling. As for my text, I've done my best to stick with British English spelling and choice of words. I've lived in the UK for more than 25 years, spent time in the US, but grew up in Germany – which probably makes me a linguistic crossbreed, too, and some Americanisms will have slipped in. No offence intended.

As a journalist[67], I ask questions, I listen and I observe until a narrative emerges. When I write, I set the scene and, in this book, I tell the story in the form of case studies, and, at times, a travelogue. The case studies provide detailed descriptions of the circumstances, the motivation that led to setting up a project, the approach chosen, the hurdles that needed to be overcome and an analysis of which aspects work and where changes and improvements are needed. There is no 'one size fits all' solution. Every project described is unique. Locations, geography, infrastructure, available capital and socio-economic environments differ, even the type of problem in need of a solution varies.

Each situation is unique but none of the solutions would have been possible, had it not been for the individuals who set out to find them and make them work. It's down to their motivation, their background, their skills and perseverance. All of the farmers, butchers and chefs I met and talked to, all the people that were crucial in making things happen, they are pioneers who saw an opportunity and grabbed it. In all this variety of circumstances, chance and individuality however is a throughline. From what these pioneers have achieved, basic principles have emerged and lessons can be drawn as to how sustainable meat production, meat processing and a local supply chain can be made to work.

Part 2

New rules – solutions from across the Channel

Chapter 4

Regulated down to the second

Herbert Siegel is an organic livestock farmer in the village of Missen in southern Germany. He began raising beef cattle in 1997. The animals were kept on grass year-round, and for slaughter, Siegel brought them to a nearby abattoir in Kempten. Things went well until, just under two years into his career as a beef farmer, he took an 18 months-old steer which he had handled and knew to be extremely docile and tame for slaughter. When he arrived at the facility after a half hour drive, the animal was visibly distressed, shaking all over and covered in sweat. When Siegel tried to lead it out of the trailer, it attacked him. "It was sheer luck that I didn't sustain any injuries," he says. The incident at the Kempten abattoir turned out to be life changing for Siegel. The experience drove an 18-year quest to find a different, more humane way to have his animals slaughtered. If deer and wild boar could be shot in the field and their meat considered to be fine for human consumption, why was the same not possible for cattle? He began discussing the issue with local authorities, animal welfare experts and veterinarians. And he wasn't alone: farmers and campaigners in other parts of Germany also pressed for changes and eventually the EU granted Germany an exemption to the existing rules, and a protocol for killing and slaughtering animals on pasture was established. In August of 2016, Herbert Siegel was able to have one of his animals legally killed and slaughtered on a pasture for the very first time.

Sigel's animals stay year round on grass.

A licensed hunter will shoot the animal within the herd.

To transport the dead animal a simple
'slaughter box' with a lid is sufficient.

Since then, the exemption has become the new rule: on 9th
September, 2021, an EU wide regulation was passed that gives a
clear legal framework for on-farm slaughter[68]. It gives individual EU
countries the option to draw up an ordinance that specifies clear
parameters for on-farm slaughter in accordance with national law. In
Germany, this type of detailed rule setting cannot be done at federal
level, it has to be left to the 16 federal states. There are differences
in administration as the organisational structure of state veterinary
services may vary, which explains slight differences in the protocols
for on-farm slaughter.

Veronika Ibrahim is a veterinarian at the Ministry for Environment,
Climate Protection, Agriculture and Consumer Protection in the state
of Hesse and she was one of the first and most active proponents of
on-farm slaughter. In November 2021, the agricultural ministries for
Hesse and neighbouring Baden Wurttemberg organised an all-day
webinar to take stock, consider the implications of the new EU
regulation and create a space for farmers, butchers, vets and other
stakeholders to share their experiences. As it turned out, in Hesse,

as in most other German states, the EU framework for on-farm slaughter allowed for the protocol to be simplified. Veronika Ibrahim walked participants through the process step by step.

An animal can be slaughtered on farm, but the carcass needs to be processed and butchered in an abattoir. The first step, therefore, is a signed agreement between farmer/owner of the animal and the processor – which sounds complicated but in effect it is a simple, short form. Next, either the farmer or the abattoir have to inform the official veterinarian at least three days in advance about the date and time of the planned on-farm slaughter. The veterinarian has to inspect the living animal and has to be present at the farm for the slaughter. A mobile slaughter unit (which can be a simple slaughter box or a more sophisticated unit mounted on the back of a small lorry) is used for the transport of the carcass to the abattoir, which has to proceed "without delay, detour and under hygienic conditions".

Only stomach and intestines may be removed on site and they have to be transported with the respective carcass to the abattoir. No further processing is allowed on farm. If the time between slaughter and arrival at the abattoir is under two hours, and climatic conditions permitting, refrigeration is not necessary. If the process takes longer than two hours, the carcass has to be transported in a refrigerated vehicle. The farmer has to inform the abattoir about the estimated time of arrival, where the carcass then has to be processed without delay. As would be the case for any animal slaughtered at an abattoir, the necessary forms have to be completed by a vet.

The mobile slaughter unit has to be licensed. Through the license, the mobile unit legally becomes an extension of a licensed abattoir. The units most commonly in use in Germany are fairly simple, suitable for slaughter and the hygienic transport of the carcass. It must be possible to completely disinfect the unit and it must be possible to close it for transport. Most units are not refrigerated and need no access to power or water. Such simple 'slaughter boxes' should not be confused with MSUs, - small mobile abattoirs which need access to a power source and water mains.

There are two types of on-farm slaughter: pasture rifle slaughter and slaughter in a barn using a bolt gun. For stunning with a bolt gun, the animal needs to be fixated which is not needed for rifle slaughter. But legally, pasture rifle slaughter at present can only be used for animals that stay on a pasture year-round.

An animal may well be killed by the rifle shot itself, but the key purpose of both, rifle and bolt gun, is to render the animal instantly unconscious. Death is defined as the animal bleeding out, which is why a trained butcher has to cut the main artery within 60 seconds from the moment the rifle or stun gun were fired. The animals can be bled out outside or inside the slaughter unit, but the blood has to be caught in a trough or shallow pan.

Ibrahim has witnessed on-farm slaughter many times and has seen what works well and what doesn't. Initially, a stunned animal was lifted by its hind leg and bled out but Ibrahim says cutting the artery while the animal is lying on its side is faster and much safer.

The person firing the rifle shot needs to be licensed to use a firearm and has to have a certificate of expertise to stun an animal. A repeater rifle should be used in case a second shot has to be fired. The shooter should be in an elevated position, for example on the back of a pickup and the distance to the animal should be no further than 10 metres. Ideally, the animal is not separated from the other animals but remains within the herd – an experienced marksman can nevertheless place a precise shot. Ibrahim showed a video of such a pasture slaughter. While the stunned animal fell to the ground, the rest of the cattle, alerted by the shot, briefly lifted their heads and then continued grazing. One cow walked a few steps to the bull that was shot, sniffed at it and then also went back to grazing.

Ibrahim recommends pasture rifle slaughter for beef cattle, water buffaloes, cattle in nature reserves and for breeds such as Longhorns or Scottish Highlands. In her opinion, the new EU framework has strengthened animal welfare considerably: transporting live animals to abattoirs can now be avoided. But for more farmers to choose on-farm slaughter, regional abattoirs have to be maintained. Reactivating small facilities and licensing new ones should be promoted. And Ibrahim reminded the audience that Hesse has a grant programme for mobile slaughter units and equipment to fixate animals for slaughter in a barn.

Chapter 5

A good life and death within the herd

Farmer Herbert Siegel had already found a way to slaughter his animals on farm, but the new EU framework and the guidelines issued by the state of Bavaria has made things easier. He has a herd of about 60 beef cattle, most were born on one of his pastures. He works with a trained hunter who knows the animals and they know him. That allows him to shoot from a very short distance. In the past, once the animal had been shot, there was narrow time frame to sling a rope around the hind leg and winch the animal up so that it could be bled out in the slaughter box. The new EU framework and the guidelines by the state of Bavaria allow for the bleeding out of the animal in a lying position. For transport to the abattoir, Siegel uses a very simple slaughter box. It is a steel box with a lid and a grid onto which the animal is placed. Bavaria allows one hour for transport without cooling; in other states farmers have two hours.

On pasture slaughter involves additional costs: for the hunter, for the vet who has to be present and for the butcher who has to sever the carotid artery of the animal. Siegel is lucky, his brother is a butcher and a community owned abattoir in the village of Seltmans is just seven kilometres away. In Bavaria in the 1990s, many butchers and small abattoirs in need of investment closed down. In Seltmans, 256 farmers from several villages came together and pledged to contribute financially to the building of a new facility, a statement of intent that meant the State of Bavaria would cover a large chunk of the overall costs.

Springtime at Hofgut Rengoldshausen.

Today, the abattoir is used regularly by 30 to 40 farmers. The only employee is a caretaker who is responsible for the upkeep of the building and for organising the schedule: farmers have to book the facility and bring their own butcher or team of butchers for slaughter and/or processing. The farmer members still pay a small annual fee, based on the number of cattle they keep. In this area, too, butchers are hard to find, but Siegel knows of several apprentices and within a 50 kilometre radius there are still six or seven abattoirs, he says.

Once the sides of beef have been hung in cold storage for two weeks and the cuts have been matured in special ripening boxes, Siegel posts online that he has meat for sale. Customers place their orders and collect them at the abattoir on a given day. Prices are 'mid-range' for organic meat, says Siegel; he wants organic meat to be affordable for anyone, not just rich customers. "The real problem is that customers only want steaks and a few other cuts. I can only sell about half of the meat that could be used." For him, keeping beef cattle and slaughtering it humanely does not create enough income, he also runs a small car tire business in order to make a living.

Vets, farmers, butchers and animal rights experts agree: killing animals on farm, on pasture, among the herd and with a rifle is the most humane form of slaughter possible. But even under the 2021 EU framework, only animals that are out on grass year-round and not housed, not even in winter, may be slaughtered using a rifle. Animals that are housed over winter, or dairy cattle which may be indoors for calving can be slaughtered on farm, but they have to be fixated and stunned with a bolt gun.

Mechthild Knösel farms near Lake Constance and spent years trying to find a solution for on-farm slaughter. Knösel keeps 'Schweizer original Braunvieh', a sturdy dual-purpose breed originating from neighbouring Switzerland – the border is just a few miles away. The farm is part of Hofgut Rengoldshausen. The estate was founded in 1932 and has been run according to biodynamic principles from the beginning. It is now owned by a trust, overseeing educational facilities, research, and biodynamic seed breeding. The land is leased to several farmers who run independent businesses, producing fruit and vegetables, eggs and chicken. Mechthild Knösel runs a dairy herd and twice a month an animal is slaughtered for meat. She farms 220 hectares, of which 100 hectares are permanent pastures. The arable land is farmed in a multi-year rotation including fodder crops such as lucerne, grains and vegetables such as carrots, potatoes, beetroot, and parsnips.

Knösel was born in Hamburg and initially wanted to train as a nurse. In the end, she felt she didn't want to spend her life caring for sick people, but rather look after healthy, living, growing things. Going into farming seemed like an obvious career choice. She did an internship at Hofgut Rengoldshausen, became an apprentice, and after several years of managing different farms, took her exam as a master farmer. She didn't necessarily intend to come back to Rengoldshausen, but in 2008 a farm position became available and she and her husband decided to take it.

In most dairy herds, conventional as well as organic, separating the calves from their mums shortly after birth is common. For Knösel, humane handling of animals begins at birth. She dreaded having to listen to newly separated cows and calves calling out to each other for hours on end, and she was one of the first farmers in Germany who decided to stop the practice. Two weeks before their due date, the pregnant cows join the mother and calf herd. In summer, the cows give birth on pasture, in winter they are moved to a spacious calving box with deep straw bedding in the barn. All new mums stay there with their calves for at least two days until they have properly bonded. Only then will they return to the mother and calf herd. The calves stay with their mums for three months and are allowed to drink as often and as much as they like. Twice a day, cows and calves return to the barn. While the cows are being milked, the group of calves waits in a special pen with deep straw bedding.

After 12 weeks the calves naturally spend most of their time with their peers. They remain in the 'kindergarden' area of the barn and enjoy time with their mums twice a day when they come in for milking.

Coming in for milking and to see the calves.

On the day we visit the farm, I wait next to the calf pen in the barn, a spacious, light, wooden structure with open sides. As the herd comes closer, the mums are calling out with calves replying. In the barn it takes a few minutes for mums and calves to find each other. Some calves greet their mums and enjoy the physical contact, others can't wait to start suckling. While some cows allow only their calf to drink and kick others away, several cows have three or four calves

Super mum.

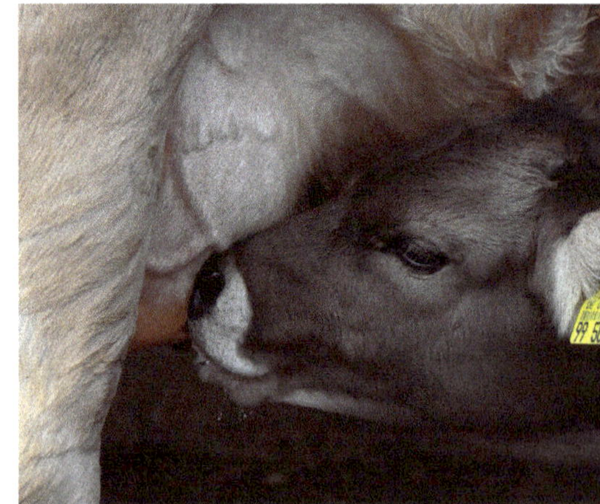

Contentment.

crowding around their udder. The cows decide how much time they want to spend with their calves, after about ten minutes the first ones begin to walk off to the milking parlour. With six stands, the milking parlour is relatively small, the cows have to wait a while until the last one leaves the stand, and the herd can return to the pasture.

The milk is not processed but sold as raw milk, mostly through the estate farm shop. The milk yield per cow is 5,000l/p.a., and each calf drinks about 1,500l. It's quite a lot, says Knösel, but she finds that the calves are much healthier and eventually the quality of meat will be better, too.

It's early evening and standing near the 'kindergarden' pen I have a magnificent view of tree-lined pastures gently sloping southward to where Lake Constance is almost visible. It's an excellent vantage point to see Knösel's husband guide the mother and calf herd up the hill to the barn for milking. An hour earlier, Knösel had taken us for a closer look. As we approached, the cows stopped grazing and watched us with interest before returning to grazing. Two young bulls dozed at the far end of the pasture. At the centre of the herd, surrounded by several calves, stood six-and-a-half-year-old Simba. He continued to watch our every move and stood guard while the cows grazed, dozed or ruminated. "He is my best member of staff," says Knösel. Some 10 years ago, she met Bud Williams, a US rancher from southern Oregon who developed 'low stress stockmanship', a method

Simba, the bull on calf care duty while the moms graze.

Low stress stockmanship means working with the animals.

to move cattle without ever shouting at them or using force. "Williams learnt to understand the animals' body language," says Knösel. "We are standing at a distance at which the animal will notice us; if we come closer, the animal will start to move." Some animals will let humans come very close before they move, others will react at a greater distance. After meeting Williams, Knösel immediately decided to learn low stress cattle handling, for her it's another important element to provide her animals with the best life possible.

From my vantage point near the barn, I am able to observe how Knösel's husband Markus slowly but surely gets the whole herd to move up the hill; while some cows are eager to get to milking, others constantly stop to graze. Knösel slowly zigzags behind them and occasionally snaps his fingers until they are on the move again, with bull Simba keeping watch from the side. "I want a homogenous herd with animals that aren't over-sensitive and can't be approached, nor should they lack respect and only react to being touched," says Mechthild Knösel as we watch the herd approach.

There is no aspect of her animals' lives Knösel has not thought about – from herd behaviour and stockmanship to the mother and calf dairy system and, in the end, humane on-farm slaughter. As a farmer, she is responsible for the animals in her care. "This responsibility begins at birth. I care for them throughout their whole lives and that means that I can't just desert them at the very end.

I do no not want to shirk this responsibility and leave it to a butcher, thinking, fine, it's your turn now, you take care of the rest." For her, on-farm slaughter was not just the right thing to do, Knösel actually learnt to use a bolt gun so that she can stun the animals herself. "To me, stunning is the crucial point," she says, "I really want the animal not to experience stress. When it is slaughtered on-farm it dies in its familiar surroundings, but at the very last moment a stranger shows up, and no matter how good the butcher is, to the animal, he is a stranger. And then the butcher makes unfamiliar moves. In the very last moments of its life, the animal will get startled and stressed, I don't want this to happen."

Legally, the qualification to stun and slaughter an animal can be obtained in a two-day course, at the end of which participants are issued with the necessary certificate. After attending the training, Knösel went to the local abattoir to learn and practice what she eventually would have to do on her farm. She doesn't say much about the training she got, but it's clear that the slaughtermen at the abattoir didn't think much of the woman farmer who came to learn about slaughter. "I knew it wouldn't be easy, but I had a very clear goal. Stunning an animal and in some cases having to use a bolt gun twice, means crossing an abyss." Knösel found a way to mentally prepare herself for using the bolt gun, to be mindful of the task, to connect with the animal she has known all its life – it is very personal – to the extent that she does not talk about it.

Knösel started researching options for on-farm slaughter in 2016. At the time, federal states could decide whether to permit online slaughter and under which conditions. Hofgut Rengoldshausen is situated in the state of Baden Württemberg. On-farm slaughter was permissible, but the rules were such that it was difficult to adhere to them in practice: within 60 seconds after the fixated animal had been stunned, its body had to be released from the fixture and a rope slung around a hind leg so that the carcass could be lifted into the mobile slaughter unit where, once the doors had been closed, a butcher was to sever the carotid artery. It was a near impossible task to complete within one minute. And why were conditions so different – while animals killed on pasture could be bled out outside, an animal stunned in a barn had to be bled out in a slaughter unit – and only once the doors had been closed? For answers, Knösel sought a meeting with officials at the ministry for agriculture in Stuttgart, only to be told that 'rules are rules'.

Undeterred, she contacted Ernst Hermann Maier, a farmer who in Germany is also known as the 'cattle whisperer'. Together with engineers from a metal construction company, he had developed the Uria MSB® II[69] slaughter box. It costs €7,000 and in Germany, it is probably the most widely used box for animals slaughtered on pasture. For the on-farm slaughter of stunned animals however, it is unsuitable, because there is no room for the butcher inside. Maier and Knösel consequently came up with the design for a prototype with doors, high enough for a person to stand in upright. Cost of production: €30,000. In the presence of officials from the ministry of agriculture, the mobile slaughter unit was first tested in December 2020. "It took us 58 seconds from stunning the animal to the butcher severing the carotid artery inside the unit with the doors closed," says Knösel, who says she still believes it was a miracle that they managed to stay under the one minute mark. In February, the unit was licenced for regular use. Then, just a few months later, in September 2021, the EU changed regulations and the requirement of bleeding out the animal in the mobile unit with the doors closed was dropped. Slaughtering has become much easier, says Knösel, the animals are far less stressed. The time she spent lobbying officials, developing the prototype and the money spent on building it, are now obsolete. If she hadn't invested so much money in her custom built MSU, she would prefer to work with the simple MSB® slaughter box because it is much easier to clean.

Knösel regularly slaughters twice a month, usually one animal, occasionally two at a time. Between 45 and 50 cows in the herd are kept for milk. All calves stay on the farm. The ones raised for meat will be slaughtered at 24 months. In an average year, 45 beef cattle and five old dairy cows will be slaughtered. The dairy cows on the farm can live to an age of 14 to 16 years and still be productive; at present, the oldest cow is 14 years old. Some are replaced at a younger age because Knösel selects animals for breeding.

The slaughter box Knösel had designed and constructed at the cost of €30,000.

Mechthild Knösel explains the yoke panel used for slaughter.

The upper part of the yoke panel.

The meat of old dairy cows can have a better quality than that of 24-months-old bulls. Knösel never castrates an animal because she will not interfere with the animal's hormonal system. Before an old dairy cow goes to slaughter, she will no longer be milked and spends six months grazing with the herd of young animals. Such older animals will have higher intramuscular fat levels which adds taste. Meat from grass-fed animals is an acquired taste, says Knösel, the animals walk all day long, and it takes a lot longer for them to reach their slaughter weight. Meat from an animal that barely moved while it was alive and was fattened on maize or soy meal will taste very different.

A few days before slaughter, the animal is brought into the barn. Knösel will move another animal next to it for company, usually a dried off cow[70]. It's not an ideal situation because the two animals don't know each other well, says Knösel. Ideally, the animal would be killed within its familiar herd environment. At the entrance to the barn, the feed barriers on either side have been modified. Feed barriers allow all animals to get their share of feed because it prevents higher ranking animals chasing lower ranking ones away. The first two feed barriers have a yoke panel so that the head of the animal can be fixated and held in place. During the last days before slaughter, Knösel trains the animal to the yoke panel by feeding it treats – parsnips are particularly popular. She also gently presses a wooden stick onto the forehead of the animal, so that it gets used to that

The yoke panel is not
set in concrete and can
be fully opened.

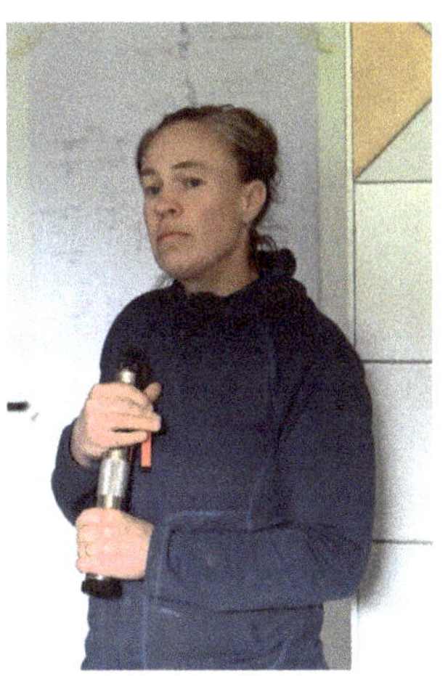

Knösel with one
of the bolt guns
she uses.

movement and the pressure. When she places the bolt gun on this
exact spot a few days later the animal will not be scared or stressed.

Animals are usually slaughtered at around 8 am. A few hours earlier,
Knösel will visit the animal one last time to say good-bye and thank it
for all it's given to her and the farm. "I have my ritual," she says curtly,
about what is clearly an important but also very private moment.

A few hours later, once the butcher and the vet have arrived and the
mobile slaughter unit is in place, the yoke panel is closed so that
the animal's head is fixated. Knösel uses the bolt gun and tests eye
reflexes and breathing to make sure the animal has been stunned
properly. She has several stun guns within reach in case one
misfires or she needs to stun the animal again, which occasionally
does happen. While the standard feed barriers are anchored in
concrete, the one modified for stunning has a metal bar at the
bottom which can be removed. Once the yoke has been released,
the animal falls to the ground, chains are slung around the hooves
and the carcass can be dragged out and slightly raised, so that the
butcher can sever the carotid artery. Once the animal has bled out,
the carcass is lifted into the mobile slaughter unit and transported to
the abattoir at the local butchers. All meats, cuts and sausages are
delivered to the farm and sold through the farm shop, meat boxes
and local butcher shops.

The system works well, says Knösel, but it is still not ideal. Because of the winter weather conditions in the area, the animals have to be housed for about four months of the year. Therefore, on-farm slaughter has to happen in the barn as only animals that are on pasture year-round can be killed within the herd. To Knösel, the most humane form of slaughter is to kill the animal in its familiar herd environment. A rifle shot is more likely to instantly kill an animal. She would, therefore, like to slaughter on pasture for eight months of the year and use the barn only when the animals are housed. Farmers in Switzerland are already allowed to work that way. The border is just 30 kilometres away and the German rule makes little sense. "To change this will be the next big fight with the ministry of agriculture and the county agricultural offices," she says. And she'll be ready if and when the rules change: she will train to use a hunting rifle and obtain a licence. Her animals will always have the best life and the best death that she can possibly give them.

The way in which animals are slaughtered on the Rengoldshausen estate may be the most humane possible, but it certainly is not the most cost efficient one. Knösel had to invest €30,000 for the mobile slaughter unit alone – she says she hasn't yet sat down to calculate the actual cost of slaughter per animal.

Klaus Bonkhoff is a third-generation butcher in North Rhine-Westphalia. Traditionally, the meat from the animals Bonkhoff slaughtered was sold through his shops, market stalls and farmers markets. In the 1990s the family specialised in halal slaughter for the growing Muslim community in the area, and in 2010 they invested in a mobile slaughter unit which was initially used mainly for emergency slaughter, for example, if an animal had broken a bone. Today, on-farm custom slaughter is one of the mainstays of the business. Bonkhoff does not work with slaughter boxes, but a bigger, more sophisticated mobile slaughter unit. The inside of a small lorry is fitted with a grid over a pan to collect the blood from the carcass, at the back wall is a winch to pull the animal inside and there is a separate compartment with a hand basin, water tank and boiler. The company runs three such units which are usually fully booked, often well in advance. Many farms will have a neck clamp which the animals are familiar with, if not, Bonkhoff's team will deliver one the night before. The next morning, bales of straw will hide the lorry and unfamiliar people such as the butcher and the vet from view while the farmer leads the animal into the neck clamp and fixates the head. It has little time to notice the butcher who places the bolt gun. The carcass is then winched into the mobile slaughter unit where the butcher severs the artery. If the animal shows signs of distress, it will not be slaughtered – the butchers just leave and come back another day.

After slaughter, the carcass is transported to the Bonkhoff's small abattoir where it is dressed. The sides of beef are aged for a fortnight, then the meat is cut and packed according to the specifications of the farmer – while some will want beef quarters, others may need portioned roasts and steaks which can be delivered straight to farm shops, packed into meat boxes or shipped to online customers.

During Bonkhoff's presentation at the on-farm slaughter webinar held in November 2021, it became very clear that humane slaughter doesn't come cheap. He charges €50 for the use of the mobile slaughter unit plus €0.80 per kilometre. The hourly rate for the butcher is €58 and the vet fees range from €30 to €104 per animal, depending on the county. How state veterinary services are organised and what they charge varies from state to state and sometimes even from county to county. Costs quickly add up. By comparison, slaughter fees at an abattoir amount to between €80 and €100.

Bonkhoff argues that veterinary costs are the decisive factor. Up to three beef cattle can be slaughtered on-farm at any one time, and that helps to bring the costs down. In his model calculation, he assumed a cattle weight of 400 kg. If a vet is present for the slaughter of one animal only, the on-farm slaughter cost is €0.69 per kilogram, on average. If three animals are slaughtered, the cost drops to €0.30 per kilogram. At present, on-farm slaughter is a niche sector. If humane slaughter is to become mainstream, it has to be economically viable. Bonkhoff argues that veterinary services for on-farm slaughter should be subsidised and free of cost to the famers. If that were the case, costs per kilo would drop to €0.44 per kilogram for a single animal and €0.17 per kilogram for three animals. The calculation makes sense, but Veronika Ibrahim, the veterinarian at the Ministry in Hesse was muted in her response – she and several vets present from other federal states were doubtful that on-farm slaughter will be subsidised in this way any time soon – if ever.

Chapter 6

When small abattoirs work best

Before you continue reading: The next few chapters deal with the actual process of slaughtering an animal. What are the necessary steps? How do slaughtermen and butchers work – in an abattoir, in an MSU? What does humane slaughter look like? In order to understand the need for small abattoirs and for slaughtering animals as close to the farm as possible, we need to know what happens and why. I tell the story, the pictures document it and....

It's not just in the UK that small abattoirs are closing down in record numbers; it's happening in Germany too. Small facilities run by cities or counties are struggling. At best, they are not making losses. Often buildings and equipment are old. An upgrade would need a huge investment and old municipal abattoirs are often centrally located.

Any cash strapped city council will jump at the idea of closing the abattoir and selling the land to a developer. In rural areas, at least in southern Germany, pretty much every small town used to have an abattoir. With commuter belts around city centres such as Munich expanding ever further, demographics in rural areas are changing, and so is the infrastructure. Supermarkets and retailers moved to greenfield sites and have rendered village shops superfluous for many customers. Traditionally, villages often had more than one butcher who slaughtered and cut meat, and farmers had a choice of whom to sell to, relationships were based on quality and trust. Today, meat, is mostly sold through supermarkets, sometimes at a meat counter, more often pre-packed and shrunk-wrapped in the cooler aisle.

In the late 1990s, farmers in Legau, a small market town (population 3,300) in the Allgäu region of southern Germany, wondered how they might reverse that trend. The Allgäu is a dairy region, and in any given year, dairy farmers usually have only a few calves, the occasional heifer, or an elderly dairy cow to sell for slaughter. As in the UK, large abattoirs process hundreds or even thousands of animals a day, they can't and won't deal with farmers delivering just one or two animals. Rainer Waizenegger, an organic dairy farmer, was one of the farmers who decided to solve the problem by setting up a slaughter facility to market the meat locally. In 1997, 65 farmers from and around Legau founded an association, the 'Bauerngemeinschaft Illerwinkel', along with a subsidiary for a butchery and a shop. Rainer Waizenegger became one of the directors.

The Legau abattoir is not sign posted. The night before our visit, we check out the building which is outside of Legau, slightly set back from the road – you do want to know where you are going at 4 am. But the next morning we have no problem finding the facility – it is the only building with lights on for miles. Patches of ground fog cover the pastures, the early autumn sky is clear, with the starlight strong enough to

The Legau abattoir.

illuminate the quiet landscape around us. Then Klaus Steinhaus, the master butcher, pulls up in his car and takes us inside where his two colleagues are already getting changed. Steinhaus is 35 years old, tall and lanky, with a big welcoming smile. His colleague, Thomas Kling is 30, broad shouldered and equally friendly. With Rainer Waizenegger, Steinhaus and Kling make up the board of directors of the farmers' association, and the limited company (100% association owned) that runs the abattoir, the butchery and the shop. Animals are slaughtered every Tuesday and Friday morning and within a week, the butchers have processed all meat into cuts, made it into charcuterie such as hams and salami, or one of the many types of German sausages to be eaten hot or cold. This Friday is typical for the butchers: five pigs and a piglet, a cow and two calves are scheduled for slaughter.

A veterinarian has inspected the animals the day before and will come to the abattoir around 6 am. The pigs have spent the night at the facility, the cow and calves will be delivered later in the morning. We step into the abattoir, a white tiled, high-ceilinged room with a gate at one end, wide enough to have a lorry dock here. To the left is an entrance for the animals to be led in. Just outside are two pens with deep straw bedding for animals to wait in and recover from their (short) journey. The pigs have already been led inside. Steinhaus starts with the piglet, he places electric paddles on the side of its head and stuns it. Suspended from a rail on the ceiling are long chains with a hook at the end. With just a few moves, Steinhaus has the piglet hooked up by one of its hind legs and with the help of the chain, it now can be easily pulled up and moved to the side. A bucket is placed under the head, before his colleague cuts the piglet's throat.[71] The other pigs hardly seem to notice what is happening. One by one, they are stunned and bled out, it takes no more than about 10 minutes until five large and one small carcass are hanging from the ceiling rail.

Next, each pig is placed in a lidded de-hairing tank for scalding, any remaining hairs are then singed off with a blowtorch. The eyes, the

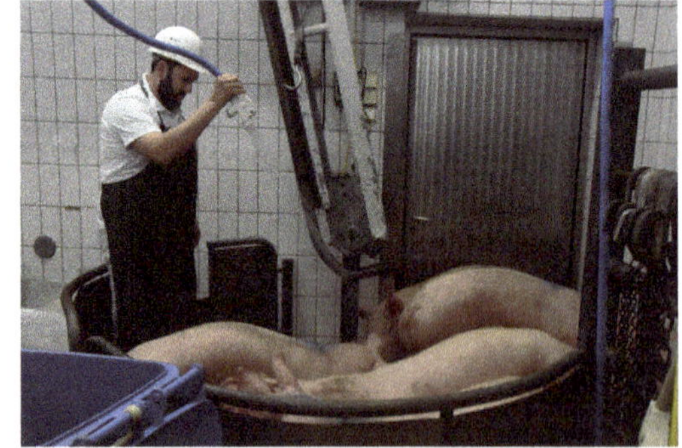

The pigs are sprayed with water which ensures that the electric tongs for stunning will work well.

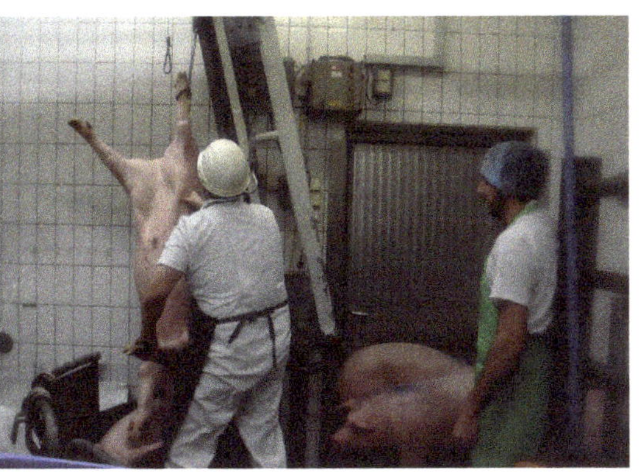

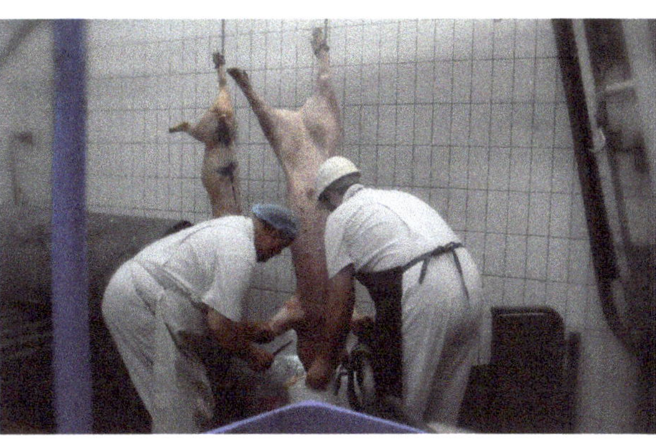

The stunned pig is lifted up...   **...the artery is cut and the animal is bled out.**

thymus gland, part of the ears and the horned part of the feet are removed with a knife. Once this is done, Steinhaus and his colleague hook up each carcass by the hind legs and move it towards the back of the room where Thomas Kling opens the belly along the midline. The intestines and some of the organs are placed in stainless steel buckets and will be disposed of later as waste. The intestines could be cleaned and used as sausage skins, but the cleaning process is long, arduous and financially not worthwhile. On this occasion, the animals' blood was considered waste too, but whenever there is demand for 'Blutwurst', the German version of black pudding, the blood is captured in a steel basin and taken to the butchery for processing. Next Kling removes tongue, lungs, heart and liver, all in one piece, and carefully places them on a wall rack where the vet will soon inspect them. He then removes fat from the underbelly, which could be made into lard for baking, but it's early September, too warm still for customers to buy lard, which is only in demand during the cold season. Before he pulls out the spinal cord, Kling makes sure there are no remnants of glands, and the gall bladder has been properly removed. Once the whole cavity is clean, Kling uses a cleaver and chainsaw to cut the carcass down the length of the spine and then moves the halves along the rails into the adjacent cool store. By the time he is done, the next pig has been dehaired and is ready for him.

Doing the work with two butchers is possible, but a team of three butchers is ideal, says Steinhaus. Who does which job is decided on the day, and if need be, positions can be swapped. Each member of the team is trained to perform every task – from killing the animal and dealing with the carcass in the abattoir, right through to preparing meat cuts, charcuterie and sausage making in the butchery.

It took just minutes for all the pigs to be slaughtered.

De-hairing with hot water.

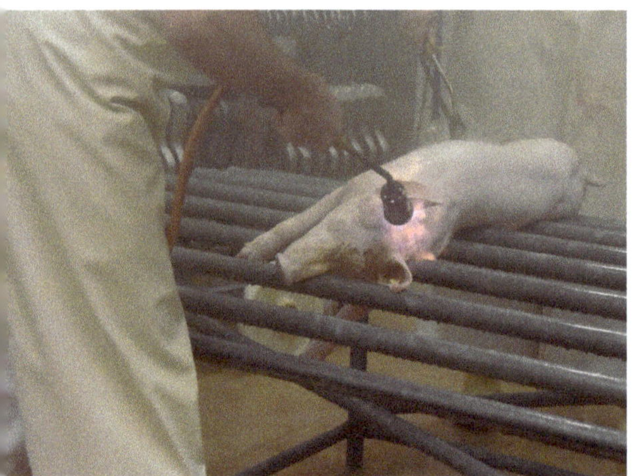

Any remaining hairs are singed off with a blowtorch.

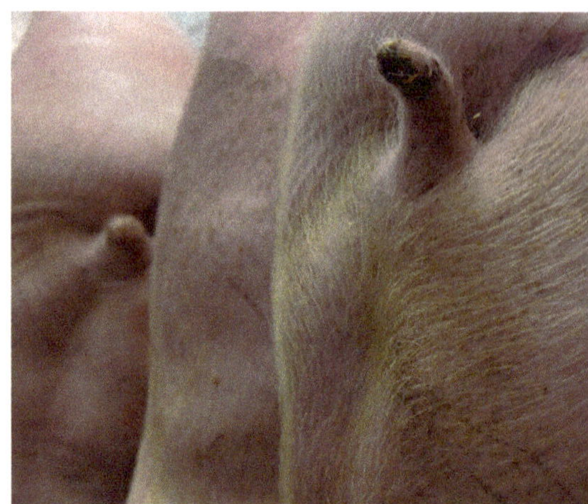

The pigs had their tails docked. Not a good sign.

Out of their 40-hour work week, the process of killing animals takes only a tiny fraction. Compare this to a big slaughter facility, where this will be done on the 'kill floor': a number of workers will do nothing but stun animals or slit their throats to bleed them out, for eight hours a day, five days a week. The carcasses will then move along an assembly line with every worker doing only one job, often reduced to a single movement. He or she will repeat that one movement every few seconds throughout the working day. Both Kling and Steinhaus say that while they deal with the carcass, they

The organs are removed
in one piece.

Heart, liver and lungs ready
for inspection by the vet.

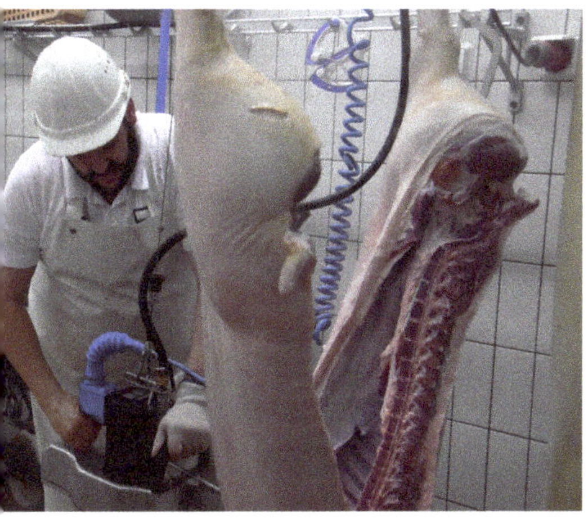

The carcass is split in
half with a saw.

get a feel for the quality of the meat and that will influence their
decision on how to use it – is it ideal for a big roast, should a piece
be cured or cooked or best used in a particular type of sausage?
"We are producing high quality food that tastes great and we do it
from start to finish," says Steinhaus.

By now it's 5.30 and a farmer has arrived, delivering a seven-year-old
dairy cow for slaughter. It's blindfolded with a large piece of colourful
cloth, and the farmer leads the cow into the abattoir, gently talking

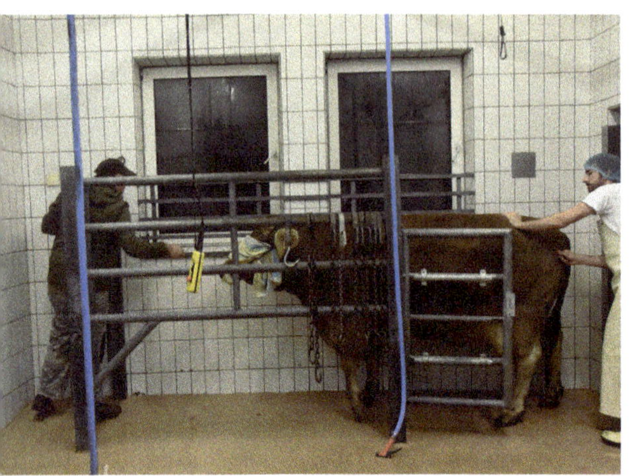

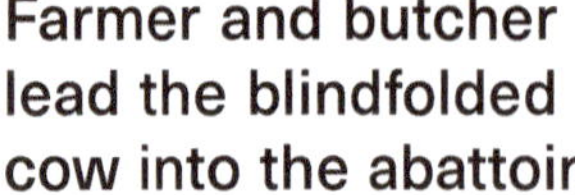

Farmer and butcher
lead the blindfolded
cow into the abattoir.

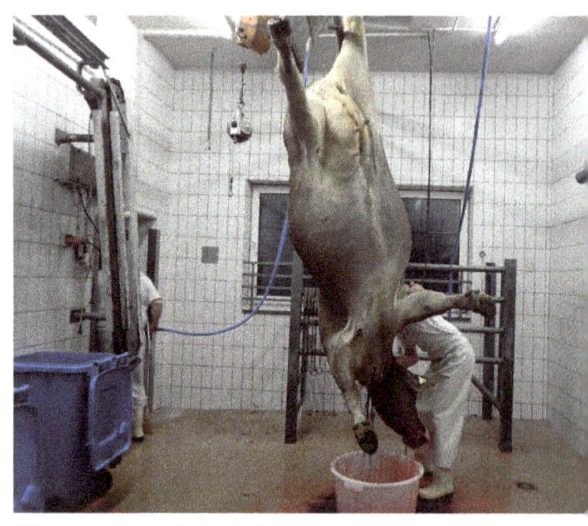

The animal is bled out...

to it. Steinhaus walks behind and as soon as the cow stands still, he moves up, places the bolt gun to its head and the animal breaks down – the whole process is over almost before I realise what's happening. Within seconds the butchers have the animal winched up by a hind leg, sever the carotid artery, and are waiting for it to bleed out. Next, the hide is removed. When the cow is cut open, its stomachs fill a whole wheel burrow, most of it is a bright green mass of semi digested grass: this cow had been grazing until just hours before she was slaughtered.

I step outside to have a chat with the farmer. Achim Ritter initially trained as a chef, but then took over the running of the family's dairy farm. He regularly brings a calf or a cow for slaughter and he sells the meat online. There had always been customers who wanted to buy 10 kg mixed meat boxes, but "since COVID the demand has gone crazy," says Ritter. He will return to the abattoir in a few days and collect the meat of his cow, cut into large chunks – he will do the actual butchering himself on the farm. The meat of a seven-year-old cow would normally be dog food, or ground for hamburgers at best. Not his cows, says Ritter, "as a chef I know how to prepare and tenderise meat. We will sell steaks and roasts from this animal and customers will come back for more".

While we are talking, another farmer has arrived and unloaded two calves which are now waiting in the pen next to the entrance of the

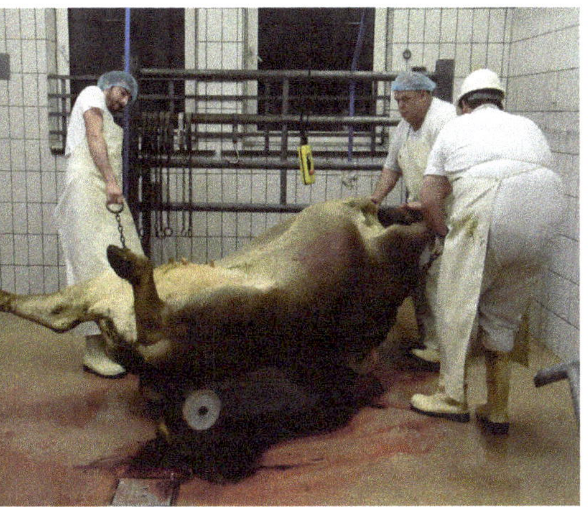

...and placed on a
sledge for de-hiding.

This cow was grazing
just hours earlier.

abattoir. Bruno Fleck has come with his two sons, Emil, aged 13 and 11-year old Pius; both wanted to be there as their favourite calves were delivered for slaughter. They take turns standing next to the pen, talking softly to the animals and stroking them. Emil's calf is called Stadi; Pius has named his 'Bauer Hermann' ('farmer Hermann'). Bruno Fleck is a Demeter certified dairy farmer, but he, too, sells meat directly. In a year, he brings 15-20 calves and two to five cows for slaughter. There are some regular customers – the local pub usually buys the meat of one calf per month. The rest sells out fast via the farm's website. Customers check in regularly, not just for meat offers. Like Mechthild Knösel, Fleck raises the calves with their mothers.

Emil and Pius are still standing next to their calves. In a few days, school will start again after the summer break, both are glad they were able to accompany their Dad. Yes, they love these two calves "but that's the way things are," says Pius, who wants to become a farmer, "we can't keep all the calves, but until we bring them here, they really had a good time".

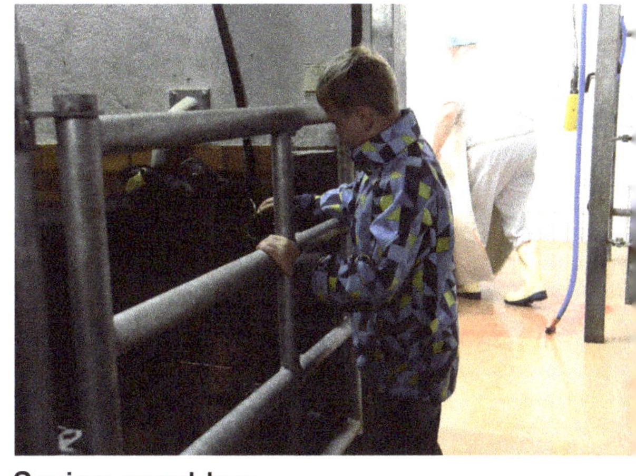

Saying good-bye.

Klaus Steinhaus comes outside, greets the Fleck family, and asks the kids about their animals. Then it's time. Emil has gone back to the car, Pius gives 'Bauer Hermann' one last hug, he waves good-bye and joins his Dad and brother.

While we were talking outside, the vet, Jenny Härle, has arrived. I find her in the cold store where she is inspecting the sides of the five pigs and the piglet. She takes samples which will be checked for trichina at the Memmingen extension of the Bavarian state veterinary department. Everything looks fine and she places a purple stamp on each side of pork – it signifies that a veterinarian has declared the meat fit for human consumption.

Härle runs a farm vet practice together with a colleague. She regularly does the checks at the Legau abattoir, which means that on Fridays and Tuesdays her workday starts before she even has breakfast with her family. Doing these routine controls does not pay well, she says. "I am absolutely in favour of small abattoirs, the animals don't need to be transported far, that and the way they are handled here, it's really important. That's why I do this, farmers need this service." It would be nice if she could do the checks later, but the samples have to be at the veterinary office by 7.30 am, or they won't be checked that day. Härle walks back into the abattoir where Thomas Kling earlier hooked up the tongue, lung, heart and liver of each pig and placed them on hooks along the wall. She points at one of the lungs, "the surface should be smooth, but can you see how rough this looks? And the heart is abnormal too. And here, this dark red bit on this lung, it indicates that the animal had pneumonia as a piglet". These findings and the fact that the pigs had docked tails raise concerns, says Härle, who will make a note in the records the state veterinary office keeps on all livestock holders. Should findings of ill health in pigs from this producer become a pattern, there will be an unannounced control visit by the state veterinary service. Would a veterinarian in a large slaughter facility be able to check as thoroughly? Härle gives me a long, raised eyebrow look and says nothing.

Checks on meat are not the only thing she does. "Especially in a small facility like this, one also needs to keep an eye on hygiene. Look here, this is the area where the carcasses are sawn in half and the head split. Sometimes the saw or the cleaver touches the floor and then you get grooves in the surface which can be a real issue

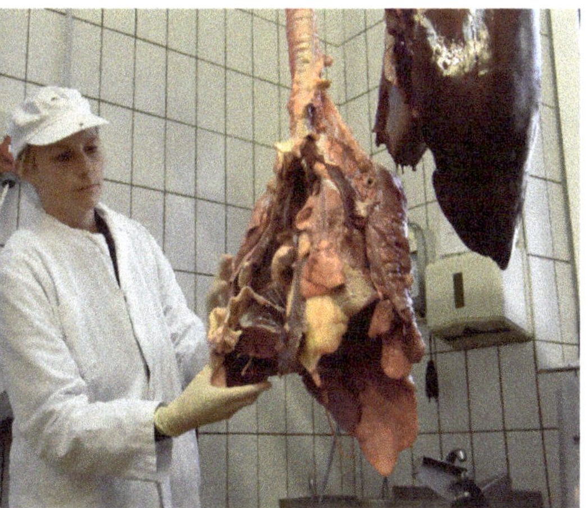

Vet Jenny Härle inspects
the inner organs...

...and filling out the paperwork.
All done before breakfast.

for hygiene. In a large facility, there is so much wear and tear the
floor probably will be resurfaced regularly anyway. In a small facility
like this, I point out things like that." And repairs are being carried
out: from where we are standing it is easy to see that part of the
floor has recently been resurfaced.

Härle is off to have breakfast with her family and we decide to have
a break, too, before we meet Steinhaus and Kling again, this time at
the butchery and shop, opposite the church and right in the centre
of Legau.

Spoilt for choice.

Germans do take
sausages seriously.

Thomas Kling and his colleagues are back from the abattoir.

Charcuterie.

When we arrive at the shop at around 9 am, three sales assistants are busy serving a steady stream of customers. The choice of sausages, ham, cold cuts, meat salads, roasts and ribs is huge. An enticing aroma hangs in the air – just as well that there's a basket with different types of rolls from the local baker for anyone feeling the desire for a freshly prepared, made to order cold or hot meat sandwich.

The shop connects to the butchery, but the entrance is at the back of the building. All of the rooms are tiled from floor to ceiling, steam hangs in the air, there are sausages being cooked in a large vat. Kling has rubbed spices into hams that are now left to cure. Steinhaus grinds meat that will be sold fresh or made into hamburger patties or meatloaves. Work in the butchery starts at 6 am, part time staff and the trainee do routine jobs before the three full-time butchers return from the abattoir and join them. "It's been five years since we last had an apprentice," says Steinhaus. In Germany, any apprentice will usually attend a trade school for one day a week, while learning on the job and on the shop floor for the rest of the time. The nearest trade school has just four apprentices who want to become butchers. "When I was an apprentice, we were 42 in my class, today, only half of them still work in the profession," says Kling. "Being a butcher has such a negative image: dirty, badly paid and you've got to get up early." The latter is true, says Kling, "but you finish work early too, particularly in summer that's brilliant". How did he decide to become a butcher? "My grandpa was a

Sausage making.

Freshly cured.

butcher, my mum worked in a restaurant next to a butcher's. It was kind of an obvious choice," says Kling, who took his butcher's exam but does not want to study to become a master. For Steinhaus it was a career choice that he took after a work experience he had to do as part of the school curriculum. "I was 14 and ended up working at a butcher's. I really liked it and there was no other profession I was particularly keen on, so I chose butchery for my apprenticeship".

Together, Steinhaus and Kling take all necessary decisions for the business. "How many animals we slaughter in a week depends on the time of year," they explain. In summer, business is better because of all the BBQs and grill parties, small festivals and events. Usually these will include a large tent where beer and food are being served, spit roast suckling pig is a particular favourite. The abattoir and the shop were profitable until the pandemic struck: customers bought more meat, but without event catering and party service overall sales went down by 20%. "Now we have parties with 30 or 40 people again, but not events with 300 or 400 guests," says Steinhaus. "We have a turnover of about a million euros, normally €80,000 come from big catering jobs. Last year we ended up with a loss of €12,000."

Another income source is contract slaughter and butchery – both the farmers we met in the morning will collect the meat from their animals and sell it directly. "We offer a bespoke service and farmers really value and appreciate that," says Steinhaus. The farmers decide whether they want to take whole sides of beef back to their farm, or individual, vacuum-packed cuts, or whether they require a full butchery service, from ground beef to curing and sausages.

The abattoir has EU and organic certifications, but the butchery is not organically certified. "Customers don't want organic meat," says Steinhaus, "it is too expensive. And they don't like the sausages because of the colour. Under organic standards we would not be allowed to use nitrate salt mixes, which means sausages might look grey rather than pink, and customers believe that's a sign of a meat product well over its sell by date – which in this instance is not the case. Another factor is the organic spices we would need for organic sausages – they are very expensive." Steinhaus and Kling are responsible for buying the animals the butchery will process for the shop as well as all ingredients needed. The animals are sourced only from farmers who are members of the farmers' association. There are some who supply relatively frequently, but others may only have an animal for slaughter every six months or once a year. In such cases, Steinhaus or Kling will visit the farm, inspect the animal on offer and only then decide whether to buy or not.

Prices for meat and sausages in the shop are similar to what customers would pay at a supermarket. The difference is the quality, says Steinhaus. Industrially produced sausages have a high profit margin: the meat used will likely be of low quality and non-meat 'fillers' will reduce the cost of the ingredients even further. The butchery in Legau is competitive on price because there is little need for storage as the products are freshly made, and customers are served by staff, which means the products don't need the kind of packing required for display in a supermarket cooling cabinet, and transport costs are minimal. Behind us a timer sounds, one of the sales ladies from the shop wants to know about a special order – we let Kling and Steinhaus get on with their work and take our leave.

In the evening, we're meeting up with Rainer Waizenegger again, who has been involved with the farmers' association Illerwinkel from the start. He fills us in on the history. The association got off to a good start and things worked really well until disaster struck in 2016: the 40-year old master butcher who had set up the butchery and built the brand – meats and sausages from the 'Bauerngemeinschaft Illerwinkel' have an excellent reputation far beyond Legau – was diagnosed with late stage cancer in April. In June, the father of three was dead. Even retelling the story five years later, Waizenegger is still visibly upset, everyone in the community was shocked and deeply saddened. To keep the business running, the farmers' association had to recruit a new butcher master – and fast. The first one left after only six weeks, the second one stayed for a year – that's how long it took to finally get him to leave. "He nearly ruined us," says Waizenegger, "suddenly, the quality of the products was awful, customers complained and stayed away. One of the sales assistants told us the butcher used only three different spices for all of the 60 types of sausage that he was supposed to make. It was a complete nightmare. In the end we cut his wages, that finally made him leave." Throughout that year, Thomas Kling continued to work alongside the butcher and tried to prevent things from going even more awry. The members of the farmers' association begged him to sit his exam for master butcher and take over the running of the abattoir and the butchery. But Kling couldn't see himself in that role. Klaus Steinhaus had done his apprenticeship under the old master butcher and later had become his deputy. But after his death, Steinhaus left to work as meat inspector. Only when it became clear that the Legau abattoir and butchery would have to close down if no one could be found to run it, did he to return. He successfully passed his master butcher exam after just 10 weeks of studying and training. Today meat and meat products from the Illertissen farmers' association are once again known for their outstanding quality.

How many animals per week are slaughtered depends on customer demand. Because farmers bring individual animals for slaughter, the system has to be extremely flexible and finely balanced, says Waizenegger. Nobody wants animals backing up on farms because there is no capacity at the abattoir. I ask about the pigs we saw in the abattoir and the health issues the vet found. He is aware of the issue. "Everybody around here does dairy, it is really hard to source sustainably raised pigs but we keep trying," – pork is needed for most types of sausage in the shop.

25 members of the farmers' association very regularly deliver animals for slaughter. The price depends on the weight of the animal and weekly listings. Initially, there were rules which didn't allow the use of antibiotics or GM feed, but there was no way to efficiently check compliance. But in the region, antibiotics are rarely used for dairy cattle anyway, says Waizenegger, and by now the dairy companies insist on the feed being GM free and the dairies do have the means to enforce that rule.

If it hadn't been for Frieda…

…and her piglets…

Any farmer from the region can use the services of the abattoir, but the fees for members are about 10% less than those for non-members. Charges depend on the weight of the animal and what additional requirements the farmer has. It's this customised service that farmers really value.

Only recently, the farmers' association was able to buy the building in which the abattoir is situated. It belonged to a farmer who emigrated to Canada and sold the building without any land at auction. The association's first bid wasn't high enough, but when the new owner lost the property in a night of high stakes poker, the farmers were able to buy it after all. Waizenegger doesn't believe that an expansion is on the cards any time soon. They would have to invest in new equipment for vacuum packaging and hire more staff – which is almost impossible.

Waizenegger believes that supply and demand are well balanced: there is enough demand for the animals the farmers want to sell for slaughter. Meat and sausages from the butchery are not only sold at the shop in Legau but supplied to 10 other shops and local supermarkets with a meat counter, as well as to several clinics and retirement homes. For now, farmers, butchers, sales' staff and customers are pretty happy with how things are in Legau.

...we would not have learnt about the Konrad butcher's and abattoir.

Had it not been for Frieda, I wouldn't have learnt about the Konrad family butcher in Lenningen, a small Swabian town some 30 kilometres southeast of Stuttgart. Frieda, a Tamworth sow, and her piglets had a wallow in their enclosure, but on a hot day in early May, she was visibly enjoying the shower her owner provided with the help of a watering can. It was a chance meeting, and the piglets still had a good few months of life ahead, but I asked the farmer where they would eventually go for slaughter? Was there a local abattoir? No, the local abattoir had closed, I was told. But there was the Konrad family butcher. You couldn't just buy excellent meat and sausages there, you could also have your animals custom butchered. And all of that in a brand-new facility.

The Konrad family agreed to our visit, even though it is an extremely busy Saturday morning in the middle of the summer BBQ season. The new facility is conveniently located at the entrance to the small town of Lenningen, next to a big ALDI supermarket parking lot. The shop faces the access road, the abattoir is at the back, and the space next to it is a big enough loading area for vans, small lorries and cattle trailers.

Gero Konrad is a second-generation butcher. His father started the business in 1985 in a remote village on a mountaintop overlooking Lenningen. Initially, Konrad senior still worked three days a week as a lorry driver, until he became a full-time butcher in 1994. Some twenty years later, shop and abattoir were in need of refurbishment and new machinery, an investment that would only make sense if his son wanted to take over the business. "My parents never pressured me," says Gero Konrad, "at school, I did an internship with a carpenter, in a metal workshop and at a butcher's shop, not my parents', a different one. And that's when I knew I wanted to become a butcher." He did his apprenticeship and went on to become a master. When his father invited him to take on the business, he decided not to invest in the old facility, but build a new one. It would serve customers wanting to buy fresh meat and sausages, and farmers from the nearby Swabian Alb region wanting to sell meat directly.

The facility cost €4 million to build.

The Swabian Alb is a high plateau with a harsh climate and mostly small scale or part time farming businesses. The area is of 'outstanding natural beauty' and an 85,000 hectare national park is recognised as a UNESCO biosphere reserve for its unique flora and fauna. "We need animals to protect this unique countryside," says Konrad, "there are a lot of small holders who keep animals, they want to direct market and they need someone who does slaughter and butchery for them to make that work. I want to help these businesses."

The new facility has been built and licensed to slaughter cattle, pigs, goats, sheep and lambs. Konrad will also get organic certification as he expects that up to a third of animals brought for slaughter, will come from certified organic farms. He expects to slaughter 40% of the animals to supply his own shop – usually that amounts to 20-30 pigs and two beef cattle a week – and 60% of his business will be custom slaughter for farmers in a 50 kilometre radius.

The abattoir is set up as to minimise stress for the animals. The improvement in meat quality is an additional benefit. Konrad owns cattle himself, so he knows how to handle them and wants every

animal to have a death that is as humane and stress free as possible. He does his utmost to ensure that the animals don't get anxious when they arrive. The farmer will be present, the animals will not be rushed, they are given time to come off the trailer and walk into the facility. He talks to them and will not allow any animal to be pushed, pulled or prodded.

Konrad slaughters up to 10 beef cattle per week. At the end of Ramadan, there is an extremely high demand for freshly slaughtered sheep and goats and during that one week, he may slaughter up to 40 animals, but never more. It is a decision he has taken despite the fact that veterinary costs would drop considerably if he were to slaughter more than 40 animals per week.

To Konrad, animal welfare and humane slaughter are more important. The amount of paperwork the butcher has to do is independent of the number of animals slaughtered. Some of the papers have to be supplied by the farmer, such as a transport permission and a certificate that the animal has not recently be treated with medication.

Konrad works together with his father, also a master butcher, four journeymen and one apprentice who is currently in his second year of training. Cattle are slaughtered on Monday and Thursday mornings, goats, lambs and suckling pigs on Tuesdays and Wednesdays, the rest of the week is needed for preparing cuts, sausage making and curing.

Konrad guarantees that each livestock holder gets back the meat from his or her animal. Almost all of the farmers want 10 kg packets containing meat, sausages and offal. Tripe, offal, pigs' heads and trotters are in high demand. "Farmers print fliers when they have them available and usually do not have enough for everyone who wants some. Many people still do traditional meat jellies, it's a very popular dish." Only 10% of his suppliers want special 'grill packets' containing only steaks, roasts and sausages.

Once the meat packets are ready for pick-up, they will be loaded into one of the butchery's refrigerated trailers. The farmer does not have to have cold storage or freezing space on the farm. He can simply let customers know that meat is available and sell directly out of the trailer. Costs for slaughter depend on the size of the animal. They range from €20 for the slaughter plus veterinary fees, €10 for cuts and sausage making, up to €250 for the slaughter of beef cattle plus up to €500 for cuts and sausage making. More and more farmers – sheep farmers in particular – use the opportunity to direct market meat, says Konrad. Dates for slaughter have to be booked well in advance.

The slaughter facility, which officially opened on 1st June 2022, took two years to build and cost €4 million. "Building a slaughter facility and butchery is way more complicated than setting up something like a car repair shop," says Konrad. "There are so many things to consider. A lot of technology, ventilation, cooling systems…all of this is extremely expensive. And getting planning permission and all the necessary permits was complicated and took a lot of time." The German government is keen to promote small, regional slaughter units and Konrad was able to receive 30% grant funding, and the state owned KfW development bank provided a loan with very favourable conditions. Nevertheless, the financial commitment is huge, but the family, including Gero Konrad's partner Vanessa, is determined to make it work – they are proud of their craft, dedicated to animal welfare and determined to provide the service small farmers and livestock owners need.

On farm killing and pasture rifle slaughter are usually discussed with cattle in mind. But what about pigs? They are extremely intelligent, genetically very close to humans, and the assumption is that they are able to experience emotions such as fear, excitement, anxiety and happiness. And (hopefully not anthropomorphizing here): watching a pig, tail curled and eyes closed, while you scratch its back – it's hard not to assume that it's happy. Butchers like Gero Konrad take as much care with pigs as they do with cattle, but wouldn't it be much better to slaughter them on farm, too?

The Hofgut Silva farm lies nestled in a narrow valley on the western outskirts of the Black Forest, not far from the Rhine Valley – cross the river and you are in Strasbourg. Even at the edges of the Black Forest, farms are often poor. The slopes are mostly too steep for arable crops, but suitable for grazing of sheep or cattle. On south

Living the good life.

facing pastures fruit trees on strong root stocks often provide an additional income, but with most fruit now grown in commercial orchards, this traditional farming system is rarely viable. The Silva farm still has pastures with old cherry and sweet chestnut trees, some surrounded by forest, which makes it an ideal environment for keeping pigs.

Judith Wolfarth and her mother Ursel bought the farm in 2008 with the intention of raising pigs. While mother Ursel moved straight away to what was then a rather dilapidated farmhouse, daughter Judith initially commuted between the farm and her job at a bank in Munich. In 2015, she quit her senior position there to become a full-time farmer. Wolfarth grew up near Karlsruhe, where her parents had a timber business. Ursel Wolfarth was always fascinated by pigs – wild boars in particular. When Judith was little, her mother volunteered to take care of an orphaned wild boar piglet named Oskar from a near-by wildlife enclosure. Oskar was joined by Rosie, the sheep and over time, more orphaned wild boar piglets spent time in Ursel Wolfarth's backyard sanctuary. Once she sold the timber business, working with pigs full-time finally became an option.

Rooting...

...taking a mud bath...

...going nose to nose...

...and enjoying being a pig.

Which pig breed to choose? Ursel and Judith Wolfarth were looking for pigs with good meat quality, suitable for pasture rearing. Their vet raised concerns over importing Black Iberian pigs from Spain. That summer, while visiting a county show in the UK, they learnt about English rare breed pigs and just knew that they would be right for the farm. They decided on Tamworth because they are robust and long legged, which is good on uneven terrain, and Berkshire, known to be excellent mothers. At present, there are about 160 pigs on the farm: 17 breeding sows (three Tamworth, eight Berkshires and five crosses) and five boars (one Tamworth, three Berkshires and one cross).

Judith Wolfarth takes us to see a group of breeding sows with their piglets. It's a five- minute drive on a gravel road snaking up a steep slope and through the forest. We leave the car in one of the hairpin bends, walk a few hundred yards and there they are: four sows and 41 two-month-old piglets. One of the sows rests in the shade of a fruit tree, another one is suckling, the third seems to be overseeing a group of piglets rooting in the soil and the fourth rests next to her hut. Wolfarth had the A-frame shelters specially designed; they are made from 4 centimetre aluminium sandwich elements, and in winter the entrance is protected by a PVC curtain. Some of the piglets race to the fence to inspect the visitors, but quickly return to playing and wallowing in the mud, the sows just briefly look in our direction. Each sow has between eight and twelve piglets which are weaned after three to four months by their mothers, who simply don't let them feed anymore. Wolfarth, then, separates the pigs depending on their sex. In Germany, male pigs normally will be castrated straight after birth to prevent them from developing a rather unpleasant smell. But Wolfarth finds that keeping them separate from the female piglets prevents that. Pigs are slaughtered at six to nine months. The pigs are fed wheat, barley, peas and beans, and because feed prices keep increasing, soymeal. The farm is not organically certified. To Wolfarth, transparency and animal welfare are more important than organic certification – even Demeter animal welfare standards are not as high as those on her farm, she says.

The initial idea was to sell a side of pork or a whole carcass to restaurants or butchers, but no one was interested. Chefs don't have the time or skills to deal with a side of pork, butchers would have to market high animal welfare meat and sausages separately and for a higher price, which didn't seem feasible.

The first pigs were ready to butcher in 2014. The nearest village just a few kilometres down the road still has a small abattoir. Things went smoothly until a new vet inspected the facility and found major issues with hygiene and animals not being stunned correctly. Wolfarth immediately stopped sending animals to this particular abattoir. For about a year, she transported her pigs to a small but very well-run abattoir in the middle of the Black Forest. But in winter, navigating narrow country roads with a trailer proved to be too stressful – for driver and pigs.

Before sending her animals to the nearest big municipal abattoir in Offenburg, about a half hour drive away, she wanted to see the facility. Like Mechthild Knösel, she decided to get a licence to slaughter and learn to stun and kill animals. Entering the abattoir for the first time was quite a shock. Slaughter starts early, so it was still dark, and it was hot, with floors slippery and stressed pigs squealing. The butcher who taught her to apply the electric tongs correctly was Ralf Wendle, a butcher master who himself was not happy with how the abattoir was run and left long before it was finally closed down in 2019.

At the time, much attention was given to on pasture slaughter of cattle and mobile units. Would a mobile slaughter unit be an option for the pigs on the Silva farm? The Wolfarths discussed the question with butcher master Ralf Wendle, who also teaches butchery at the local college for vocational training in nearby Kehl. The slaughter process for pigs is different than that for cattle. While there is a time window of at least two hours to get a beef carcass to an abattoir for de-hiding, gutting and butchering rigor mortis in a pig sets in after just 20 minutes. Once they have been dehaired through scalding, they have to be opened up and the entrails removed almost immediately. A mobile slaughter unit suitable for pigs, therefore, has to have access to water, a waste-water tank, high voltage currency, as well as a cooling unit for transport. Custom building such a unit would cost between €250,000 and €300,000. Killing one or two pigs on farm would not make financial sense. The unit would need to be operated by two butchers and be in use five days a week, to be viable. Add to that the considerable bureaucratic challenges: a mobile unit would have to operate across county lines and get licensed in each county. Vets also do not work across county lines. For the Wolfarths, it became clear very quickly that coordinating a weekly schedule for a MSU would be a full-time job and simply not feasible. But Ralf Wendle believes that in future, running a mobile slaughter unit might be a good business model for two young, independent butchers.

Judith Wolfarth and her mum were undeterred, their pigs would have a good death – even if that meant building a slaughter facility on the farm. It took three and a half years to get the necessary permits, and a million euros to build the small, but state of the art abattoir and butchery, which includes sausage making equipment and a smoking and ripening chamber.

Wolfarth butchers once every two weeks. Michi, the farm hand, moves the animals selected for slaughter to a pasture near the abattoir three days prior. We have come to Hofgut Silva on a slaughter day. Judith Wolfarth has manoeuvred the trailer close to the pasture. As we approach, Michi calls out to the pigs and shakes the feed bucket. The animals obviously know him and run to the gate. A trail of feed leads the three pigs and a five-year-old breeding sow up the short ramp and into the trailer. Then it is a 150 metre drive to the abattoir. Butcher Ralf Wendle, his colleague and the vet are waiting at the entrance as Michi opens the back of the trailer and the first pig, curiously sniffing the air, comes out. Michi and Judith Wolfarth talk to it in a low voice and guide it towards the building. Once again, a little feed helps. As it's inspecting the entrance Wendle attaches the electric prongs, the stunned pig is slung up by its hindleg and moved into the abattoir where the butchers sever the main artery to bleed the animal out. The other pigs take no notice. "Animals don't have a concept of death," says Wolfarth. A few minutes later the next pig walks towards the abattoir. Only the sow, five-year-old Mausel, is hesitant. She doesn't like to walk across the block paving. Michi realises this makes her uneasy and spreads some awning like a carpet for the sow to walk across. Mausel has had four litters and normally would have stayed in the herd. But the last time she was with a boar, she seemed to be in pain. An hour later, when the vet examines her organs, he finds large cysts in one of her ovaries. Wolfarth is glad she took the right decision, there wouldn't have been a treatment to cure Mausel.

Getting out of the trailer.

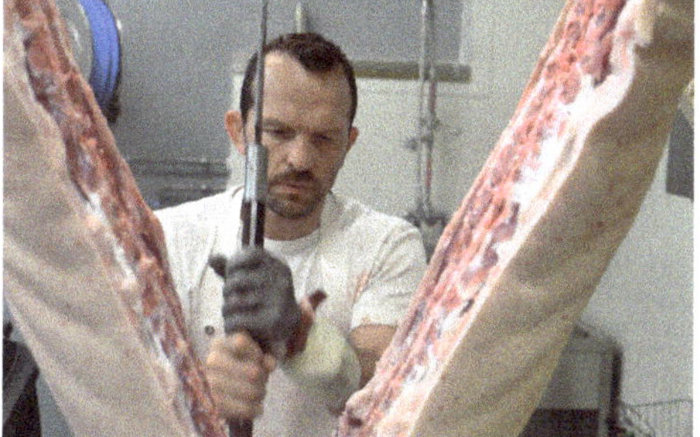

No fitness studio needed...

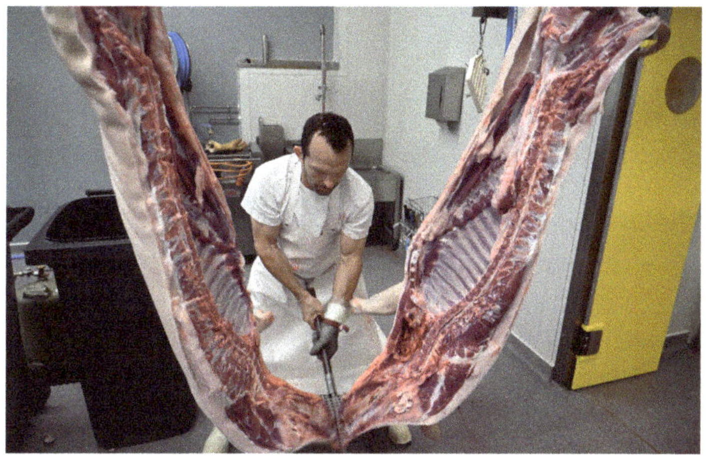

...this takes precision, muscle power and lots of skill.

The two butchers work with quiet concentration. Once Ralf Wendle has removed the organs, he makes sure he has a firm stance and with three or four precisely placed cuts of the meat clever, he splits the carcass in halves. Once the vet has inspected the animal's organs, the sides of pork are moved into cold storage.

Wendle will return the following day to cut the meat, ready to pack and sell to customers. On the third day, yet another butcher makes a variety of sausages, some cooked, others cooked and smoked, while Judith Wolfarth cures the hams that will not be airdried. For 10 days,

all hams will be kept in one of
the temperature and moisture
controlled chambers in the
butchery. Some will then be
smoked, others will be left
to ripen for two years, on or
off the bone. Do we want to
take a look? Of course we do.
We leave the butchery with
its tiled walls and gleaming
stainless-steel equipment and

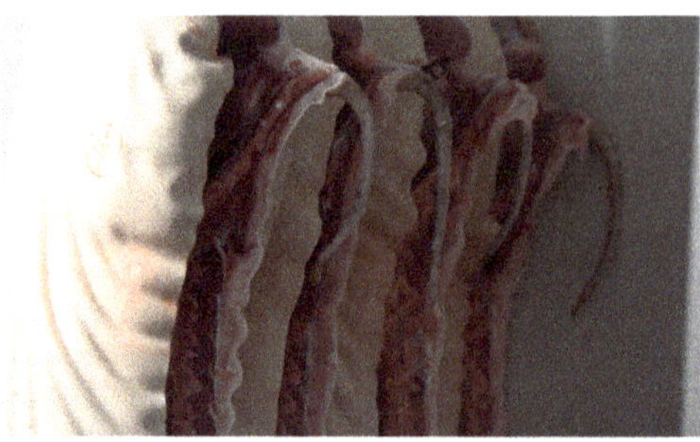

In the cooler. Note the intact tails.

follow Wolfrath to the barn, an old structure with thick stone walls,
partly built into the mountain slope. She opens the heavy wooden
door to a small room fitted into the corner of the building and we get
a glimpse of rows of sausages and several dozen hams hanging side
by side, each with a date label. A stream of cool, fragrant, slightly
briny air hits us, the thick walls muffle all outside sounds and if it
weren't for the modern and functional steel racks, we just might have
stepped back in time to a previous century.

The Wolfarths direct market most of their products. Customers can
order packets of meat, sausages and ham at a price of €25 per
kilo and arrange a date and time for pick-up or have it sent by mail.
They also deliver to restaurants, hotels and shops with high end
delicatessen. On the events schedule are two farm tours per month
which include taste testing the products, and every two years a big
farm festival with a five-course menu cooked by a Michelin star chef.

For the abattoir to become profitable, she would need to slaughter
360 to 500 animals per year, at present it's 180. Because the facility
is considered to be part of their farm, it cannot be used for the
slaughter of animals from neighbouring farms. "We keep as many pigs
as is appropriate for the 18 hectares of land we have," says Wolfarth.
Under current regulations, she is permitted however to slaughter
animals in her abattoir if they were raised on land she owns, even if
it were further afield. Many farms in this part of the Black Forest are
no longer financially viable. It is easy to find work in nearby cities
and older farmers with marginal land on steep slopes increasingly
decide to sell. The Wolfarths are thinking about buying up such
land and leasing it to young farmers wanting to raise pigs to their
strict animal welfare standards. They are looking into finding a legal
framework which would allow for those animals to be slaughtered at

the Silva farm. "This is a project, and our farm is the model," says Wolfarth. She and her mother want to show that it is possible to give pigs a good life and a good death and eventually be profitable. Both are convinced they will achieve their goal but to get there "you really have to have deep pockets". As the Wolfarths discovered, just because the abattoir is small and local does not mean animal welfare standards are good and the slaughter process is humane.

In Germany too, conditions in meat processing plants became headline news during the pandemic. Tönnies is Germany's biggest meat producer with an annual turnover of over €7.3 billion (£6.2 billion). In 2020, 1,500 employees tested positive in a single plant. Consumers learnt that slaughtering animals is the job of workers, mostly from Eastern Europe, recruited by contractors, working with little protection, for minimum wages, often living in overcrowded dormitories. The work takes place on assembly lines in huge plants because today, three German companies slaughter 58% of all pigs and 48% of all cattle. 'When did slaughter become an industry?', became a much asked question. As big slaughter facilities remained closed for weeks due to COVID, and slaughter-ready animals backed up on farms, everyone began to wonder what had become of all the small and regional abattoirs that used to exist.

With elections due in several of Germany's federal states, politicians were keen to introduce popular policies, such as initiatives for small, local and regional slaughter facilities, which, they argued, would provide better working conditions and higher animal welfare standards. Lower Saxony's agricultural secretary went as far as demanding a small abattoir for every county. First off the mark, with a concrete plan and a budget of €11 million (£9.4 million) was Baden Württemberg[72]. The modernisation of existing facilities would be funded up to 40%. Funding is also available for new slaughter units owned and run by producer associations, with animal welfare being the highest priority – an important caveat, as over the years inspections of small slaughter facilities have uncovered minor and major regulatory offenses. In November 2020, undercover videos by animal rights activists led to the closure of a small facility in Baden Württemberg. In 2019, the Albert Schweizer Foundation published a report[73] that examined the slaughter of 678 animals at small abattoirs in the state of Hesse between 2014 and 2017 and found one or more minor and/or major breeches of regulation in 44% of cases. Most had to do with ill or untrained staff working under time pressure.

On its own, the provision of funding for new facilities or the modernisation of old ones is insufficient, concluded the Unabhängige Bauernstimme[74], a monthly publication for organic and family farms. There also needs to be the political will to create a level playing field that does not financially disadvantage small slaughter facilities. The paper found that Tönnies and the second biggest meat processor, Westfleisch, pay less than two euros (£1.70) per pig. For independent butchers who slaughter only a few animals per week the fee can be as much as €12 (£10.23). Bauernstimme calculated what that means: Tönnies slaughters 16 million pigs annually, but the company pays less than €30 million in fees. Westfleisch kills seven million pigs per year and pays less than €12 million. If, for example, the inspection costs were fixed at six euros per animal, the cost to Tönnies would be about €96 million and €42 million for Westfleisch. "It is easy to see what it would mean for a company that posts about €10 million in profits annually, if it were treated like your average butcher: the company would be bankrupt," writes the Bauernstimme.

Slaughter waste disposal, too, gets more expensive the fewer animals are processed. The county around the city of Cloppenburg in northern Germany is known as a centre for the factory farming of pigs (and chickens). The fee for the disposal of slaughter waste from one pig is €40.62, getting rid of the slaughter waste of a thousand pigs and the charges drop to €0.56 per animal. As long as the costs in fees for small abattoirs is so much higher than for industrial scale meat processors, small outfits will have to make savings elsewhere if they want to stay in business.

One option is working with cheaper, less skilled or unskilled labour. That will save money, but ill paid staff working under time pressures will often have animals handled roughly and stressed unnecessarily which, in turn, increases the probability of insufficient stunning. It seems that we truly get what we pay for: cheap meat will come from animals raised in confinement, slaughtered in huge processing plants by badly paid and often unskilled labour. Meat from animals that had a good life, suffered little or no stress on the short journey to an abattoir or artisan butcher, where master butchers and well-trained staff are in charge of slaughter, and processing will cost a lot more.

Chapter 7

Butcher training, humane slaughter and meat quality

"A good butcher likes animals and knows how to handle them," says Ralf Wendle when we meet him in a classroom of the college for vocational training in Kehl, where he teaches. "A butcher holding a piece of meat and making sausage must remember that this once was a living being." Wendle likes working with metal and would have gone into engineering had it not been for his dad who worked part-time and, as a side business, custom butchered animals for neighbouring farms. Son Ralf helped out and picked up the craft. Skills that proved very useful for the passionate hunter who not only dressed his game, but that of his fellow hunters, too. Wendle became a master butcher and did a degree in business administration. He worked with various independent butchers as well as in abattoirs and is now teaching at the college in Kehl, one of Baden Württemberg's five vocational training centres for butchers. His teaching obligations still leave him time to take on jobs as a custom butcher, such as the work he does for the Wolfarths.

Like all apprenticeship programmes in Germany, butcher training too is closely regulated and combines work at the apprentice's employer with training at a college. Depending on the college, apprentices either attend one or two days a week, or, as in Kehl, they stay at the college full time for three-week intensive block courses, four times a year. If, at the end of three years, they pass both theoretical and practical exams, they will be butcher journeyman. It takes an additional 320 hours of learning and practical training to take the exam for master butcher. Over the past 10 years, classes have been small with three to ten apprentices per year. The Baden Württemberg ministry of education threatens to close the course if there aren't at least eight apprentices in every group. Employers have increasing difficulties in finding trained staff, says Wendle, and closing colleges would compound the problem, as apprentices would have to travel further. The butcheries licensed to teach and the college work together closely. Wendle lets them know when he will teach a particular skill, such as curing, and employers will try to give the apprentice the opportunity to practice. The close cooperation is one of the reasons why hardly any apprentice quits, says Wendle.

The curriculum is expansive and includes physiology, food hygiene and practical skills from preparing cuts of meat to making sausages, curing and smoking. Wendle has his students experiment and develop a good sensory awareness: what does fresh meat taste like? how do taste and consistency change when beef has been hung and aged? how does stress affect meat quality? why is it necessary to process a pig within three hours after slaughter, if one wants to avoid using nitrate salt mixes? In addition to the basic training, the apprentices have to choose two of four special topics: slaughter, meat sales, regional meat products and ready-to-cook meat preparation.

Fabienne, the plaster cast cow with removable muscle groups and organs.

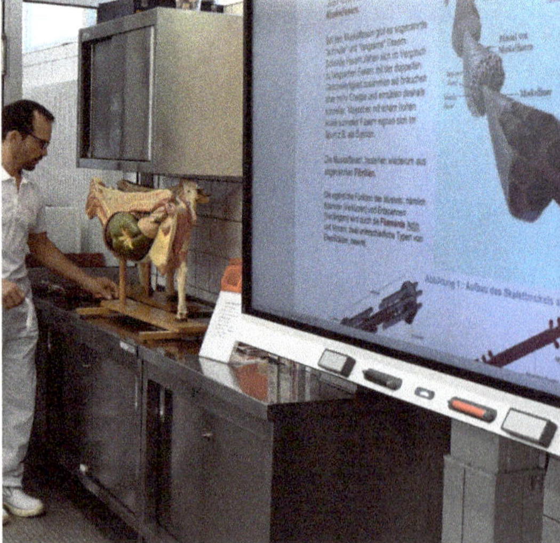

Fabienne will not be replaced by high tech touch screens.

The following morning, we are back at the college to meet the third-year apprentices. For practical training there is a butchery, fitted with any imaginable type of sausage making, meat cutting or curing implement. Next to it is the classroom with a huge, interactive, multifunctional screen. But no high-tech device can replace Fabienne, the plaster cast cow with removable muscle groups and organs. There is a difference between handling something and seeing it on screen. Six students are ready to talk to us. They all agree that there are good reasons to become a butcher: it's a sought-after profession with high job security, wages are decent and better technology has made it less physically demanding, it's varied and creative. At 17, Fabio is the youngest. He began his apprenticeship straight after finishing school. His grandpa was a butcher, the butcher's shop was on the ground floor of the house he grew up in. It closed years ago, but most of the machines are still there. Maybe he will be able to reopen his grandpa's shop one of these days. He likes meat and wants to know where it's from, how it's been prepared. He has chosen the slaughter module because he believes that killing an animal is something a butcher must be able to do. Aron is 19. He comes from a family of butchers and hopes to take over the family business one day. He finds theoretical learning a challenge but loves the physical work and being able to create something with his hands.

The butchery behind the classroom is fully equipped.

The other apprentices are older and becoming a butcher is a second career. Danny worked as a supply chain manager when he decided to retrain as a butcher at age 30. He spent time with family in Texas and it was there that he saw a documentary about artisan butchers, which fascinated him. He works at a supermarket meat counter and would like to go into sourcing and procurement. Eventually, he would like to work in the food industry or for one of the big supermarket chains. He can't imagine slaughtering an animal, but "you don't have to do that as a butcher, if you don't want to".

Elias' parents have a butcher shop and a small abattoir. They buy animals from local farmers. After his A levels, Elias did several internships – working as a pipe fitter, as an architectural draughtsman and as an industrial management assistant – but he doesn't want to "sit in an office and stare at a screen all day". He, too, wants to work with his hands. He has chosen the slaughter module, but eventually wants to work in sales because it is paid better. Once he has passed his exams he wants to study food technology and business administration to qualify for a senior management position.

For 22-year-old Alexander, the apprenticeship as butcher is also a second qualification. He is a trained retail salesman and works in a big supermarket. A job rotation scheme had him spend time at the meat counter. He absolutely loved the work and decided to do an apprenticeship. He, too, does not want to slaughter animals and thinks he wouldn't be able to, but he definitely wants to make and sell meat products.

At age 42, Michal is the oldest of the apprentices and 16 years ago he moved to Germany from Romania. For six years, he worked as a cutter at Tönnies, at piece rate wages. Then he met his partner and moved to a village where he found a job at a local butcher. One day his boss told him that his job would be better paid if he had a formal qualification. Because of his long experience, he would only have to do a short, two-year apprenticeship, and there was government funding available for him to get better pay from the start. For Michal, it was an offer he couldn't refuse.

Ralf Wendle is visibly proud of his third-year students and pretty sure that in two weeks' time, they will all pass their exams. Each of them already had several job offers – the choice is theirs. And butchery is not just for men. "We usually have at least one female apprentice," says Wendle. And the women do well. "One was a trained tax consultant, a job she found utterly boring. When she became a butcher apprentice, her dad threw her out of the house. Today she is a master butcher."

Ralf Wendle enjoys combining his teaching job with regular work at the farm abattoir, which gives him an opportunity to practice and hone his skills. And Judith Wolfarth consulted with him during the planning and building phase of the abattoir, so Wendle pretty much got to design his own workspace. But, for the Wolfarths profitability is still a long way off.

What if a company got behind the idea of an on-farm abattoir? Lindner Group is a family-owned construction company with headquarters in Bavaria and 7,500 employees worldwide. It is a highly specialised construction company for everything from insulation and clean room technology to acoustic construction – one of their projects was the Elbphilharmonie concert hall in Hamburg. The founder, Hans Lindner, is now 82, and the company and its subsidiaries are mostly run by his four daughters and their spouses. Despite Lindner Group being a global business, the company maintains strong ties with the local community. Among a number of social initiatives facilitated through the Lindner Trust is a multigenerational living facility, which is home to 250 people. And Hans Lindner does follow his dreams – there is now also a company-owned brewery. It was over a birthday dinner in early 2016 that the Lindner family got talking not just about good beer, but also about good food. Meals in the company's canteen were heavily subsidised, but what were they serving? Shouldn't the food be organic? And if so, why not branch out into agriculture and produce their own organic fruit, vegetables and – this is Bavaria – meat?

The man who tells me this story not only attended that birthday party, Josef Straubinger is now the director of Lindner Land und Forstwirtschaft Ltd (farm and forest business) and in charge of the farm.

It's a beautiful summer morning that makes us forget the late evening drive from Munich airport through thunderstorms and torrential rain, with so many lightning strikes that the otherwise idyllic Bavarian countryside felt like the stage set for a horror movie.

There is a lot for visitors to do and see.

Piglet youth gang.

At the start of our visit, Straubinger suggests we climb the spiral stairs to the top of an old grain silo. From here, we have an excellent view of the farm – the old cow barn had a vaulted ceiling and was converted into a restaurant and a farm shop; the extension towards the small country lane houses the farm office, packing and storing facilities and the butchery. The mobile slaughter unit is parked right outside. Across the yard is a modern barn and machine shed. Straubinger explains the layout of the farmland: 45 hectares are

in the direct vicinity, 85 hectares are scattered within easy reach.
70 hectares are certified organic and the rest is under conversion.
A third of the land is permanent pasture, two thirds are used for
growing wheat, barley, triticale, oats, peas and field beans in a long
rotation, and clover grass leys improve soil fertility.

Back down on the ground, our tour of the farm starts with the pigs, a
mix of 723 pink Schweizer Edelschwein landrace and black and white
Schwäbisch Hallische pigs. Both breeds are known for their excellent
meat quality. The 55 breeding sows are kept on pasture in groups
of eight to ten, and with access to a community A-frame shelter.
Straubinger has trees planted, but until they are big enough, the
pigs have to make do with the shade provided by recycled pieces of
tarpaulin and sheets of fibre glass supported by metal poles. A week
before their due time, the sows are moved to a fenced farrowing
pasture with an individual A-frame hut, including water access and
a private wallow. The sows are artificially impregnated – the job of
the farm's Duroc-Pietrain boar only is to get them into the mood.

Group housing.

A mud bath is happiness.

Farrowing is staggered so that a new group of sows gives birth
40 days after the previous one. Schwäbisch Hallische pigs farrow
after exactly 116 days, which means the night from Wednesday into
Thursday. Young sows may need assistance, but mostly Straubinger
and the stockman are greeted in the morning by eight or nine healthy
piglets. The biggest problem are older sows who decide not to use
the huts to farrow in, but build a nest on pasture where the newly
born piglets are at risk of getting too cold. After 40-45 days, the

piglets from three sows are naturally weaned to form a "kindergarten group" which is kept in a small fenced area with shelter. Once the piglets reach a weight of 20-25 kilos, they get access to about 1 hectare of pasture.

Josef Straubinger lives in the nearby village and enjoys his walk to work, especially very early in the morning when, in summer, the air is still cool. But his favourite time is in winter after the first snowfall – to be there at sunrise and see the piglets come out of their shelter to explore that strange cold white stuff on the ground is a sheer delight, he says. Straubinger comes from a farming family. He was always interested in organic farming, and when his older brother developed severe allergies and could not take over the farm, Straubinger would have stepped in if his father had agreed to a transition to organic agriculture. But his dad wasn't to be convinced. Straubinger contemplated a career in landscaping, did a degree in glass construction and wound up getting a job with the Lindner Group, working far from home in northern Germany for many years. But he always wanted to return to Bavaria and the dream of running a farm never quite disappeared.

We walk back to the farmyard and to the cattle shelter, a simple wooden construction, open on two sides towards the pasture, the grazing area for the heifers and bulls designated for slaughter. The cow and calf herd, a mix of two rare breeds, Murnau-Werdenfelser and Original Braunvieh, stays on one of the farm's more remote grazing areas. Cattle are kept in groups of six or seven and slaughtered when they are about 24 months old. The open shelter is also a feeding area, there is always fresh hay to be had and the animals are free to walk in and out. The feeding barrier right next to the door to the farmyard is only accessible if the animal walks a few yards through a narrow walkway. This barrier has a yoke panel, similar to the one we have seen on Mechthild Knösel's farm.

The butchers come to the shelter regularly, so the animals are familiar with them. On the day of the slaughter, the animal that chooses to come to the yoke panel, voluntarily and free of fear – the butchers must be able to touch it – will be the one that is slaughtered.

The farm has a mobile slaughter unit. On the day of our visit, it was parked next to the abattoir and was just being cleaned. The unit was developed in cooperation with a trailer manufacturer and a company for butchery supplies. The electricity needed for things such as lights, the winch, the electric tongs and the moving the hydraulic feet that stabilise the unit, can be supplied by an on-board battery. But on the farm, there is access to water and electricity on every pasture because cables and pipes for water have been installed at a depth of 1 metre. In this part of Bavaria, winters still can be really cold and once temperatures drop

Josef Straubinger explains where the animal is stunned and the carcass can be removed through a gate similar to a garage door.

below 5°C, the drinking water will be warmed slightly by a heating coil installed in each of the troughs. For slaughter, access to water and electricity is, of course, handy too.

The slaughter unit, which cost just under €60,000 to develop, build and kit out, has a 200 litre water tank. The wastewater from the MSU and the abattoir are fed through an on-farm separation filter. Most of it will be clean enough to discharge into the environment, but a small remaining amount is picked up once a month by a special company equipped to dispose of it safely.

With a height of four meters, the unit is licensed to be on the road and suitable for the slaughter of animals including beef and pigs. Normally the slaughter weight of a pig is about 100 kilos and a much smaller unit would suffice. But for sausages and some other meat products, the butchers want pigs with quite a bit of body fat: the meat of old breeding sows is ideal for making salami, and with a slaughter weight of 160-170 kilos, the carcass of these pigs is almost as long as that of beef cattle. The unit is fitted with a cooling system, "just to be on the safe side". On the farm, it is needed only on extremely hot days.

For the slaughter of beef cattle, the MSU will be parked with the back to the cattle shelter. One of the butchers stuns the animal with the bolt gun, while the second butcher stands outside and removes a wood panel next to the animal. A wedge on the opposite wall causes it to slide towards the opening, which makes it easy to sling up the hind leg, winch the body up and then sever the carotid artery. By law the butchers have 60 seconds for the whole process, but they usually take no more than 40 to 45 seconds, says Straubinger. The abattoir is just across the yard – the distance from the pasture is so short, both pigs and cattle could easily be made to walk it. To slaughter them in the MSU is a very conscious choice, says Straubinger, walking the animals to the abattoir would be less humane because for the last minutes of their lives they would have to be taken out of their familiar environment, causing unnecessary stress and anxiety.

Pigs are slaughtered on pasture, usually when they are seven or eight months old. Each pasture has a feeder that is moved frequently, so that the animals learn to look for it. For slaughter, the feeder is placed at the back of the unit. The animal that walks in first will be the first one to be slaughtered: the neck of the animal is wetted, it is stunned with electric tongs, and bled out. When the butchers need pigs' blood for sausage, a hollow sticking knife is used, and the blood collected. Once the carcass is bled out, it will be hung in the back of the unit behind a partition. The feeder is set out again and the next animal walks in – the process is repeated until the butchers have the number of pigs they need for that day.

Per week, the butchers slaughter eight to ten pigs and one beef cattle. A vet comes the night before to inspect the living animals and the next day for the organ inspection. By law they don't have to be present for the slaughter itself, but they do come for the occasional spot check.

The abattoir was fitted into an existing farm building at a cost of roughly €250,000. A new building would have been considerably more expensive – the Wolfarths spent four times as much. But repurposing an old building has drawbacks too: the butchery section is quite cramped and it's sometimes difficult for the butchers not to be in each other's way.

How many animals will be slaughtered is decided the week before. Lindner Group canteens, the kitchen of the multigenerational living facility, the on-farm restaurant, nearby hotel kitchens and restaurants, as well as online customers have until Wednesday to place an order for the following week. Once the information is in, the schedule for slaughter and the butchery can be made. On average, around 4.5 tonnes of meat per week are needed. The number of pigs that can be slaughtered on any given day is limited because the window for warm processing – and thus avoiding the use of phosphate – is just three hours. Meat, sausages and dry cured products such as ham or Bresaola are on sale in the farm shop and on several farmers' markets. Initially, the farm had stands on markets as far away as Munich. But that meant 18-hour work days for staff and that went somewhat against the farm ethos, says Straubinger. The team wants sales to be mostly regional and encourage customers to visit the farm to learn about animal welfare, organic food and sustainable production.

More prep to be done...

...ripening to perfection.

What about profitability? The Lindner Group canteens and other Lindner Group customers don't pay a market price for the meat they order, otherwise the meat business would already be profitable. But the farm has only been operational since the beginning of 2017 and the meat side was hit hard by restaurant closures during the pandemic. The online shop was launched in 2020. Initially, Straubinger wasn't sure whether online sales would fit into the concept of slow, natural and regional produce, but visitors to the farm wanted to enjoy at home, what they had tasted on the farm. At present, 50 to 60 high value parcels per week are shipped to customers. The online business is already profitable, and Straubinger is optimistic that the rest will follow suit. With over 100 different types of sausages and cuts available year- round, and with additional seasonal offers, the product range is huge. Restaurants have started to put offal on the menu. Knowing the demand a week ahead and working with a slaughter plan allows the team to make use of all parts of the animal and minimise waste.

The farm has five full time employees. The shop, the restaurant and online sales have created quite a number of additional jobs, some of them part time, which is very important to people in this rural and rather remote region. The abattoir and butchery work with three full time staff (a butcher master and two journeymen), and there is now also a butcher apprentice.

I was able to catch up with the journeymen during their lunch break. Vincent Eckel mostly worked for small butcheries slaughtering no more than 30 pigs a week. He comes from a farm family, his father raised pigs conventionally which Eckel really disliked, but he couldn't convince his father to transition to an organic system. In Germany, the Schweissfurth Trust promotes organic agriculture, education and the production of organic food. The trust also funds a training centre for different professions in agriculture and the food industry, including butchers and bakers.

In his mid-thirties, Eckel decided he finally wanted to know more about organic farming and visited the training centre in Hermannsdorf, just east of Munich. He immediately got a job offer and stayed for six years. For personal reasons (his partner lives in the region), he quit his job at the Schweissfuth Trust training centre and applied at the Lindner Group farm. He says he'd never work in a large slaughter facility, "They don't do artisanal butchery, you don't need any skills to work there. There is no more humane way of killing an animal than on this farm."

To work as a butcher within an organic farming system takes a completely different mindset, says Eckel. "You have to research and revive old practices, use proper recipes for making sausages without additives. We have developed our own recipes and create the spice mixes ourselves." He says their products are distinct in flavour and aroma, "You know that it's one of our products when you taste it."

His colleague, Konrad Römer, also grew up on a farm. He started a degree course in agriculture but found that book learning wasn't right for him, he wanted to work with his hands and with animals. After a summer internship on the Lindner farm, he became the farm's first butcher apprentice. In 2021, he passed his exams and stayed on as butcher journeyman. "I am proud to be a butcher," says Römer, "all my friends think it is cool, too. I can always bring really good meat to any party."

As in Baden Württemberg, in Bavaria, too, few young people choose a career as a butcher; the college covering training for eastern Bavaria had just 13 apprentices in 2021. Many become butcher apprentices because they have no idea what else to do, says Römer. For him, this small on-farm abattoir is a very special place to work in, he would not have wanted to do his apprenticeship anywhere else. And certainly not at one of the big supermarket chains; in his opinion, butchery there has nothing to do with artisanal skills and the core of what he considers butchery to be.

The purpose of the Lindner Group farm is to grow and produce local food for local people, and to be part of a regional food system, says Joseph Straubinger. By now, the farm also provides apprenticeships and other training opportunities in agriculture, in the restaurant, the shop and in online sales, too. It's a major contribution to reviving the rural economy. If people can find a job locally, they won't be forced to leave the region in search of work, others may be able to return or simply choose to move to a rural area. The farm has become a model business which the Lindner Group Trust is now mirroring in Romania, says Straubinger. The Lindner Group connection with Romania began when the company was tasked to renovate a large Deutsche Bank building. When Hans Lindner visited the site, he found that all of the office furniture – desks, chairs and storage units – were still in far too good a condition to be thrown on a skip. In cooperation with a charity, the furniture was delivered to kindergartens and schools in Romania where it was desperately needed. Now many kids who sat at those desks are leaving school with little prospect of finding a job; like their parents, many will leave to find work in other EU countries.

To create job opportunities, the Lindner Group Trust bought a 400 hectare farm from an Austrian farmer who decided to move back home. The farm is under organic conversion. There is arable land, pasture, an orchard and a large area for vegetable growing. The herd of 200 Angus beef cattle is thriving. By the end of 2022, the abattoir will be fully functional. Apprenticeship programmes are up and running: for farmers, vegetable growers, butchers, bakers as well as in sales and food production – the farm has a brand for its line of jellies, jams and pickles. The curriculum for the apprenticeships combines practical work with classroom learning. Joseph Straubinger has been instrumental in setting up the farm and he regularly travels to Romania. He believes that after their training, the apprentices will be well qualified to farm or start a business in their country.

Why do farmers and butchers want to change the system? Why are they willing to invest huge amounts of money into mobile slaughter units or even an abattoir? What motivates them to battle with bureaucrats, sometimes for years? Why do vets – literally – go the extra mile to make on-farm slaughter possible? Whomever we talked to, the answer was always the same: it's about animal welfare. Farmers who have looked after animals from birth don't want to see them scared, frightened, hungry or mistreated in the last hours of their lives. And butchers want to honour the animals by producing good food. As one put it: when you hold a piece of meat, never forget that it came from a living animal. None of them could imagine work as a slaughterman in a facility run by one of the big meat companies where, during an eight-hour shift, they would have to do nothing but stun and kill animals.

Only in passing would farmers and butchers say, 'Oh, sure, the meat quality of an animal that was slaughtered humanely is much better, too'. The great taste of a product isn't just important for customer satisfaction, it has financial implications too: the meat of an older animal, such as the seven-year-old dairy cow slaughtered at the Legau abattoir, can provide meat that is as good as that of younger animal raised for meat. And whether a farmer can sell an animal for steaks or just for dog food makes a big difference financially, too.

The stress level of an animal at the moment of slaughter has an effect on the meat quality. That's not just hear-say or the result of some random taste tests, it's a scientific fact. One of the AHDB's (Agriculture and Horticulture Development Board) websites explains why this is: "During the natural conversion of muscle to meat, the pH will fall from around 7 to 5.45.7. When an animal has experienced a period of chronic stress (typically 24–48 hours before slaughter), the pH does not fall to the optimum level and stays higher, resulting in meat which is compromised. A higher pH will cause the meat to appear darker in colour, firmer to touch and hold fluid within it. For this reason, it is referred to as dark, firm and dry (DFD), also known as dark cutting beef (DCB). In addition to its abnormal colour, DFD meat has reduced keeping qualities and is prone to bacterial spoilage. It is thought that there may be an improved tenderness to DFD beef. However, this benefit does not outweigh the negative impacts on both animal welfare and meat quality.

Acute or shorter periods of stress (around 45 minutes pre-slaughter) can cause the pH to drop very quickly following slaughter. This will typically give an optimum ultimate pH but will have implications on the meat quality. A quick pH fall can cause the meat to appear lighter in colour, have a softer texture and feel wet as it has a reduced capacity to hold fluid. This is referred to as pale, soft and exudative (PSE). Whilst this is not commonly seen in lamb, it can be seen in beef and more frequently pork."[75]

A number of scientific studies have shown the correlation between physical parameters such as the blood cortisol level of an animal and meat quality. In her PhD thesis[76], the veterinary Hanna Sophie Wullinger-Reber looked at mobile slaughter of free-range pigs and aspects of animal welfare, meat quality and food safety. Measuring lactate and cortisol levels, she was able to show that the slaughter was nearly completely stress free. She concludes: "The most important prerequisites for a largely stress-free slaughtering with the mobile slaughter trailer are a sufficient number of trained staff, an early adaptation of the pigs to the trailer and the avoidance of individual separations of the animals immediately before the electronic stunner...In summary, the results of this study show that mobile slaughtering of free-range pigs fulfills all requirements for animal welfare, animal protection, food safety, meat hygiene and meat quality from a veterinary and hygienic point of view and can also be legally compliant."[77]

Meat quality may be one beneficial side effect to animal welfare, another one is the benefit of meat, grass-fed beef in particular, for our health. In her book, *Defending Beef*[78], Nicolette Hahn Niman makes the argument for eating beef: it is an incredibly nutrient dense food. Small quantities provide us with amounts of proteins, vitamins and trace elements that we would struggle to consume with a purely plant based diet. Even a vegetarian diet can lack vital nutrients. Hahn Niman was a vegetarian for over three decades and initially decided to eat meat again for health reasons, though these days she enjoys grass-fed beef for the taste of it, too.

In their book, *What Your Food Ate*[79], authors David Montgomery and Anne Biklé, make the case not just for grass-fed beef, but for organic dairy as well. Organic dairy cows have to have access to grass at least for part of the year. Unlike beef cattle, it may not be feasible for dairy cattle to be grazed year round for climatic reasons. But the slogan 'you are what you eat' not only applies to humans, but to cattle too.

"Ruminal microbiota will ferment whatever a ruminant eats, creating two main by-products – gases and fatty acids. A ruminant's diet affects both gas production and the types of fats that become part of its milk and meat[80]," say Montgomery and Biklé. They go on to explain how the types of fat differ, the importance of polyunsaturated fats such as omega-3 and omega-6 and why their ratio matters for the nutritional value of meat and dairy, which are our only source of beneficial fats such as CLA[81].

Fred Provenza, professor emeritus in Behavioral Ecology at Utah State University says that animals have an 'inner wisdom' that lets them make smart choices, seeking out exactly the type of plant or other food that supplies them with the nutrients they need in a particular season, condition or phase of their life. "Plants tell stories about the relationships of herbivores, omnivores, and carnivores with landscapes. Plants are the glue that links soil with herbivores, omnivores, and carnivores – below and aboveground[82]". Provenza describes how livestock are linked to the landscape they grow up in and the feed it provided. "Moving animals to unfamiliar settings severs links with landscapes and explains why animals placed in unfamiliar social and physical settings suffer more from predation, malnutrition, overingesting, poisonous plants, and poor reproductive performance compared with animals born and reared in the environments[83]".

Provenza, too, believes that a better distinction has to be made between the nutritional value of grain fed and grass-fed beef. "In America, retail sales of grass-fed beef reached $272 million in 2016, up from $17 million in 2012. That amounts to 4 percent of all beef; the market for grass-fed beef grew at 100 percent annually during that time period. With growing interest in consuming grass-fed meat and dairy, people should assess the degree to which grass-fed meat and dairy products are better for health. The lack of research on how herbivore diets affect the flavor and quality of meat and dairy for human consumption reflects the fact that researchers, livestock producers, and consumers are just beginning to appreciate the value of soil health and plant diversity in the diets of herbivores and the implications for the health of soil, plants, herbivores, and humans[84]".

Shinn and Pledger cite a 2021 report, 'Health-Promoting Phytonutrients Are Higher in Grass-Fed Meat and Milk': "The authors assert that the nutrition in the meat of pasture-fattened cattle can't be matched by corn-fed, feedlot beef. Furthermore, they state that the meat from pastured animals offers nutrients from plants that are not cultivated as vegetables for human diets, and therefore are not likely to be available in diets that exclude animal products. According to their report, phytonutrients – the health-promoting, natural chemicals in plants – become concentrated in the meat and milk of grazing ruminants[85]".

"The United States is the globe's top beef producer. We can, and should, lead the world in forging ways of raising cattle that reverse environmental degradation and produce healthy, nutrient-rich foods[86]," concludes Hahn Niman.

In Britain, according to the Sheffield University land cover atlas[87], "agricultural land used for animal grazing and growing crops makes up approximately 56% of the country". The bigger part, 28.7%, is pastures, an additional 5.8% are natural grasslands. Grasslands are a unique natural capital. To preserve it, we need ruminants, cattle and sheep, to graze it. And for the system to work, we need farmers to keep livestock and for us to eat meat and we need an infrastructure – small abattoirs and slaughter trailers – that allow for the humane, stress free slaughter of animals.

Part 3

Across the Atlantic: prairies and 'niche meat'

Chapter 8

Prairies, bison and the Dust Bowl

It's a sunny but crisp autumn day, under the deep, blue, cloudless sky, the seemingly endless expanse of the prairie has a stark, forlorn beauty. A strong, cold wind sweeps across the land, bending the dry brittle grasses into rippling waves. The prairie grasslands in the Badlands National Park in South Dakota are at the transition point between tall-grass prairies, where more rainfall is available, and grasses can grow to shoulder height, and short-grass prairies, common in the dry regions of the western US.

The vegetation indicates where there is more or less water.

Prairies once stretched from what today is Canada to Mexico. Now, just 3% of that grassland remain. And according to the online publication Modern Farmer the area continues to shrink. "In 2021, roughly 1.6 million acres of grasslands across the Canadian and US Great Plains were plowed over. Since 2012, we've lost nearly 32 million acres, some to development, some to the expansion of farming"[88].

Over millions of years grasslands co-evolved with ruminants. Under these prairies and steppes, grazed by aurochs and bison, rich, black, fertile soils (chernozem) developed. Eventually, much of the grassland was ploughed up and turned into arable land. What we know today as the 'breadbaskets of the world', from the US corn belt, the arable areas in South America, to the wheat fields in the Ukraine, used to be prairies, pampas and steppes. "The enormous potential of grazing used to be and still is the key to basic resource of world nutrition: the trinity of biological diversity, ecological quality of water and soil fertility – the latter associated with a reduction of climate relevant emissions.

Grassland are stunningly beautiful.

The old assumption that soils have a saturation point for the additional sequestration of carbon is still prevalent in science. In fact, the potential for humus growth is unlimited"[89], says Anita Idel. "While grasses benefit from sustainable grazing, other plants have evolved in a different way. Many invest a lot of energy to grow thorns, spikes or to produce tannins to defend themselves against being eaten and the loss of biomass. Others escape being grazed by growing close to the ground. (...) Grasses have adapted so perfectly to herbivores that they in fact benefit from being grazed. They don't spend energy to defend themselves, instead they over-compensate the loss of biomass. For all other plants, a loss of biomass is a disturbance". Grasses only suffer if they are over- or under-grazed.

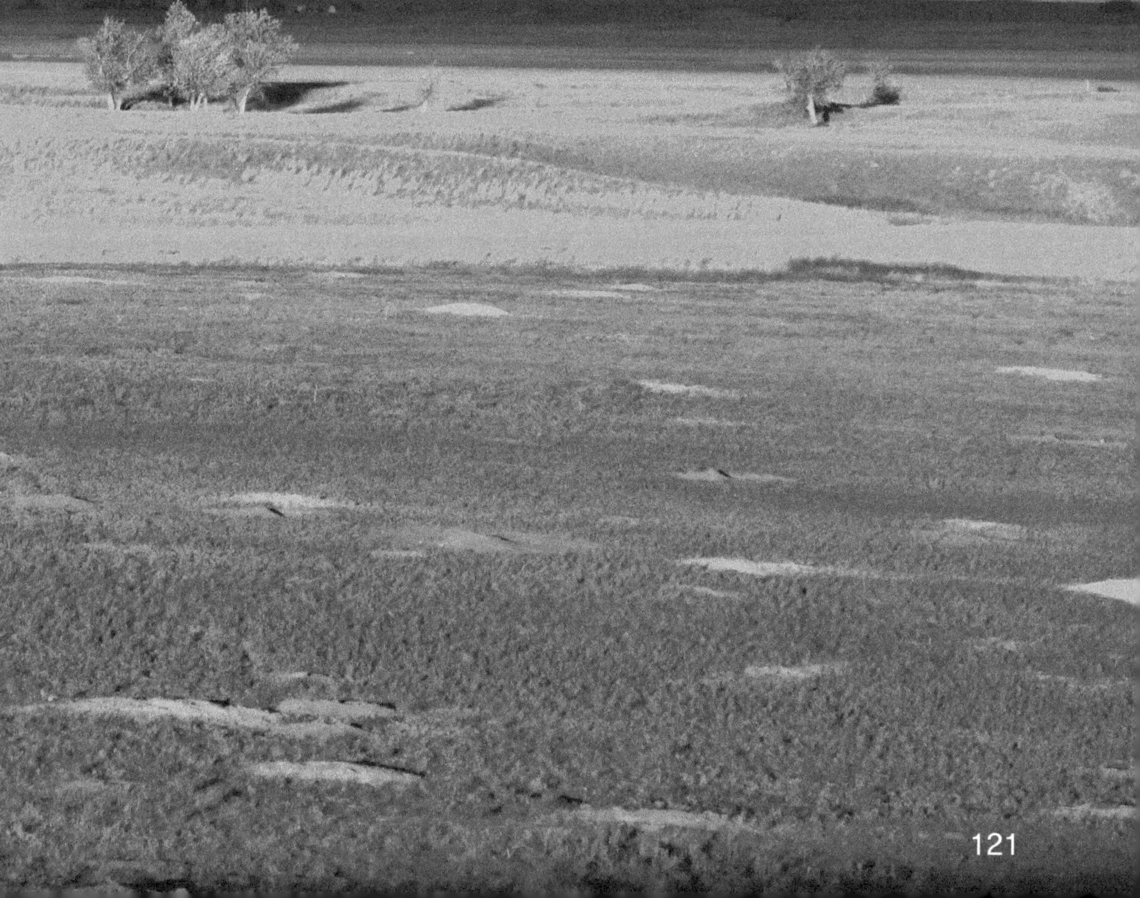

Grassland sequesters 50% more carbon than forest soils, says Idel. Trees sequester carbon above ground, it goes into growing the tree's trunk and branches. Grasses do the opposite: "Compared to other plants and trees, their (fine) roots provide grasses with the special ability to produce more biomass below ground than above. What is a root today will become humus tomorrow. That's why when it comes to promoting soil fertility, on many soils nothing beats sustainable grazing - coupled with rest phases. The root shoot ratio in grasses varies between 2:1 and 20:1, favouring root mass[90]", writes Idel, who has been fighting the bad reputation of cattle as 'climate killers' for years and believes a change of perspective is needed. "Sustainable grazing by cattle and other ruminants stimulates root growth and thereby humus building and carbon storage in the soil. In humus, more than half the content is carbon – the C from the CO_2 in the atmosphere. Each additional ton of humus in the soil absorbs about 1.8 tons of CO_2 from the atmosphere. Contrary to common assumption, cattle can help to limit climate change. If we let them...[91]." In other words: there are sound environmental reasons for farmers to keep ruminants such as cattle and sheep. If permanent grassland is grazed sustainably, which means neither over- nor under-grazed, ruminants not only produce healthy food for us, such as milk and meat, they help mitigate the climate crisis.

Even the Bighorn sheep enjoy the view.

To maintain what is left of the prairies, the Badlands National Park authorities decided in the early 1960s to reintroduce bison. Today, the park is home to a herd of about 1,200. Once a year these animals are corralled and get a health check. The herd cannot be allowed to grow because it would lead to overgrazing. As there are no predators that could keep the herd size in check, surplus animals are distributed to other parks and donated to Indian tribes.

Tell-tale signs...

...of the bison that escaped the yearly count.

When we visited the Badlands in early October 2022, the bison had just been corralled in an area in the north-western most corner of the park. We were about half a kilometre away when the car was suddenly enveloped in a dust cloud so thick that it was near impossible to make out the road. Finally, we spotted the herd, gathered near the fence, heads lowered against the wind, a beautiful, almost otherworldly sight. How had a clear but windy day without warning turned into a dust storm? As we approached the fence, the explanation became obvious: the bison had been in the corral for less than two days and even though the fenced area was huge, the grass had been trampled down so hard that in large parts the soil lay exposed. On that windy day, a dust storm was the result, carrying away the topsoil. Suddenly, it was not hard to imagine the 1930s, the Dustbowl era: millions of acres of prairie had been ploughed up to plant wheat. A few years later when the bottom dropped out of the world wheat market, those fields lay deserted. Then a drought struck and with near constant high winds blowing on the High Plains, enormous dust clouds formed that carried prairie soil as far as the East Coast and even out into the Atlantic.

Within seconds everything disappears in the dust clouds.

The dust cloud is as thick as fog.

It may be a well-known fact, but it cannot be stated often enough: globally, permanent grasslands such as prairies, act as carbon sinks, and to maintain them they need to be grazed.

In recent years, more attention has been paid to the fate of native Americans – from historic injustices and cruelty to the plight of their lives on reservations today. For thousands of years, the lives of Plains Indians were intertwined with that of buffalos, a topic film maker Ken Burns explored in the 2023 PBS series 'The American Buffalo', "tracing the animal's evolution, its significance to the Indigenous people and landscape of the Great Plains, its near extinction, and the efforts to bring the magnificent mammals back from the brink"[92].

In March 2023, Grist, a non-profit media organisation, reported that Interior Secretary Deb Haaland established a federal working group for bison restoration: "The group, which will be composed of representatives from five federal agencies and one tribal leader, is charged with creating a 'shared stewardship plan' by the end of the year to increase bison populations on lands managed by the federal government and tribal nations. 'The American bison is inextricably intertwined with Indigenous culture, grassland ecology, and American history,' Haaland said in a statement. Her agency also

**It must have been stunning to see a
herd of bison on the open prairie**

announced some $25 million from President Joe Biden's landmark
climate spending bill for bison conservation. (...) To bring more bison
back, the Interior Department's order calls for conservation based
on the best available science as well as Indigenous knowledge and
management techniques. It's one of several recent actions from the
Biden administration to prioritize Indigenous culture and expertise,
including new consultation requirements for federal agencies whose
policies could impact tribes."[93]

A few months later, in October, the U.S. Department of Agriculture
(USDA) announced a pilot project that will give three Tribal Nations,
the Cheyenne River Sioux Tribe, Standing Rock Sioux Tribe, and
Lower Brule Sioux Tribe, the opportunity to buy locally produced
bison meat through the Food Distribution Program on Indian
Reservations (FDPIR), a government food assistance programme.
"The pilot will look at changes to how USDA purchases bison to
better support buying the meat from local, small, and mid-sized
bison herd managers and delivering it directly to their local tribal
communities. (...) These local purchases will reduce the time and
distance the meat travels to the consumer, increase economic
development market opportunities for tribal and local bison
operations, and provide high quality, nutritious foods for nutrition

assistance programs. 'USDA recognizes the role its purchasing power can play in providing access for smaller, local, and tribal producers,' said Agriculture Secretary Tom Vilsack. 'We're pleased to take this step forward toward offering locally raised bison directly to the tribal communities where those herds are located.'"[94]

Five contracts were awarded to native producers and are worth over $281,000. "These contracts are the latest USDA's efforts to incorporate traditional foods into Indigenous diets across the country; in some ways, this pilot program is the most significant and groundbreaking. Though bison was already in the program, modifications were made to allow the agency to source from smaller, largely Native-run purveyors who raise their meat more in line with longstanding cultural customs[95]", says Sarah Ventiera from Ambrook Research[96]. If the pilot is successful, it could be scaled up and extended to other tribes.

Programmes like the USDA bison purchasing scheme are a real boost for producers and the local economy. But for any system to work, cattle ranchers have to get their animals to market and that's never been easy. The meat industry in the US has a chequered history, to say the least. Its origins are the cattle towns that sprang up at railheads and along railway lines. The rail infrastructure made it possible to transport animals for slaughter to destinations as far as Chicago. In his 1906 novel, *The Jungle*, Upton Sinclair portrayed the abysmal working and living conditions, as well as the appalling lack of food hygiene, in the Chicago meat packing industry. The book led to a public outcry over the lack of food safety; a congressional hearing followed, which led to the Federal Meat Inspection Act and the establishing of a federal agency that later became the Food and Drug Administration (FDA).

After World War II, agriculture underwent huge changes that impacted meat production and processing in numerous ways. The driving factor was the increase in corn yields, say Shinn and Pledger: "The problem of what to do with it all led to the proliferation of feedlots that fatten cattle on corn and other grains as grain became – and still is – relatively inexpensive. Cattle from many farms were aggregated in feedlots to spend the final months of their lives eating corn-base rations instead of grass. In the 1960s, with the shift from railroad transport of meat to refrigerated truck transport, slaughter plants and processing plants were built near the feedlots. Then slaughter and processing began to take place under one roof, and over the years those roofs became larger[97]". Today, the meat industry is highly consolidated and that has a huge impact on farmers and ranchers, the majority of whom have become price takers. "Fifty years ago, ranchers got over 60 cents of every dollar a consumer spent on beef, compared to about 39 cents today. According to the USDA, between 1977 and 1997 the percentage of cattle slaughtered in small packing plants (defined as killing fewer than half a million cattle a year) dropped from 84 percent to 20 percent. Between 1998 and 2018, the number of US cattle producers dropped by 175,000, from 875,000 to 700,000".

Sinclair describes the meatpacking district in Chicago in the early 20th century, the dirt, grime, grease, stink and noise of the stockyards where thousands of workers toiled to slaughter and process tens of thousands of animals. Today, slaughter facilities are 'hidden in plain sight', says Timothy Pachirat. For five months, Pachirat, a political scientist, worked undercover in a meat processing plant in Omaha, Nebraska. Unless one knows the address, Pachirat finds that the best way to locate a slaughter facility, is to follow a cattle truck. "Facing outward, this industrialized slaughterhouse blends seamlessly into the landscape of generic business parks ubiquitous to Everyplace, U.S.A., in the early twenty-first century. This city block-wide windowless box of gray corrugated steel set on a shoulder-high slab of concrete, topped by massive ceiling fans, surrounded by black asphalt, a chain linked fence, and guard huts, and fronted by a modern office complex of glass and aluminum presents only the generic face of mass production. The building materials, size, structure, and layout of the slaughterhouse could pass for the community college to the south, the tool factory to the east, or the pet-supply store to the north[98]".

For an outsider, to get access to an industrial facility is not easy, for a slaughter facility it is near impossible – unless you go undercover. Pachirat says he "entered the kill floor to provide an account of contemporary industrialized slaughter". He gives a detailed account of the layout of the kill floor, the workflow and the 121(!) different job functions. Killing the animal and processing the carcass has been broken down into over a hundred single tasks, from inserting a hook to cutting off the left ear and nostril, another worker does the same with the right ear and nostril, from head trimmer to liver harvester or offal hanger.

Pachirat started work as a liver hanger and worked as cattle driver before he was eventually promoted to quality control. Most tasks, including that of a liver hanger, can be performed by unskilled workers who will be paid minimum wage. The work is physically demanding and repetitive, the pace is dictated by the speed of the line, workers often need to stand in one place for long periods of time. The turnover is high, it's a job one does until something better can be found. As unpleasant as it may be to cut off ears or clean guts, it's a job that is done to a carcass. It's something completely different to work on 'the outside', driving the cattle from the pens up a ramp to what is called the 'knock box' where the animal will be fixated and stunned. "Running up the serpentine with swinging heads, the cattle are no more than a few inches away from us, separated only by the torso-high sides of the chute. Some poke their noses up over the chute wall to sniff at our arms and stomachs. I can run a bare hand over their smooth, wet noses, a millisecond of charged, unmediated physical contact. At close range, even caked in feces and vomit, the creatures are magnificent, awe-inspiring. Some are muscular and powerful, their horns sharp and strong. Others are soft and velvety, their coats sleek and sensuous. Thick eyelashes are raised to reveal bulging eyeballs with whites visible beneath darkly colored irises[99]", writes Pachirat. He calculates that during one work day, on average, 2,500 animals are killed at this plant, one 'every twelve seconds'.

The job of the men driving the cattle through the squeeze chute and up the serpentine is to keep "a steady stream of raw material entering the plant"[100]. The workers are to use plastic paddles to move the animals on but most rely on an electric prod. "Without the electric prods, the momentum of the line of animals is sufficient to move the cattle through the opening in the slaughterhouse wall into the knocking box, but not at the pace the chute workers want. When shocked, the animals jump into the box, moving the line more quickly and reducing the probability of an animal's balking and holding up the line behind it[101]". USDA inspectors are present at the slaughter facility throughout the day, but workers are on the lookout and alert their colleagues through hand signals and whistles.

Out of the 121 people working on the 'kill floor', only one, the knocker, actually does the killing. Pachirat describes the process in great detail, how, in a good scenario, the cow's eyes glaze over, indicating that the animal has been rendered unconscious.

"Sometimes the power, angle, or location of the steel bolt shot is insufficient to render the cow unconscious, and it will bleed profusely and thrash about widely while the knower tries to shoot it again[102]".

During the four days Pachirat worked at the chute, he says he drove no fewer than 6,000 animals to the knocking box. Being a knocker "man that will mess you up" a co-worker has told him, "man, that's killing...that shit will fuck you up for real". The sentence resonates deeply with him after just four days: "'Fucked up' is exactly how I feel; it is how I would describe many of the chute workers, and it captures the rawness and violence of the perpetual confrontation between the living animals and the men driving them, myself included"[103].

The RSPCA[104] defines 'humane killing' as: "when an animal is either killed instantly or rendered insensible until death ensues, without pain, suffering or distress"[105]. Pachirat's account of his time working at the meat processing plant in Omaha adds another meaning to this definition: it's inhumane work. Slaughtering 2,500 animals in an eight hour day amounts to killing on an industrial scale. Of the 121 workers, only a few deal with the living animal and only one does the actual killing. These workers are doing an inhumane job. They have to leave their empathy in the changing room together with their street clothes. But even when they do their best to insulate themselves against what they see and what they do, without a doubt "it messes them up".

Humane slaughter therefore is a question not just of animal welfare, but of human welfare, too. The examples from Germany have shown how skilled and experienced craft butchers work in small abattoirs. The act of killing takes a fraction of their work hours. It leaves them time to meet the farmers, take care of the living animals to make sure they don't get unduly stressed or anxious and to produce high quality food. Or, as the butcher master and trainer Ralf Wendle put it: "A butcher holding a piece of meat and making sausage must remember that this once was a living being."

Upton Sinclair's book did not cause an outcry over the living and working conditions of the labourers in the Chicago Stock Yards, the main concern was the lack of food safety. Today, the US Department of Agriculture, USDA, and the FDA are responsible for making sure food is safe to eat and it's not always easy to know which regulator regulates what. (The FDA, for example, is responsible for whole eggs in their shell; the USDA looks at egg products.). Regulating meat is mainly the job of the USDA. One of its agencies is the Food Safety and Inspection Service (FSIS). In 1996, FSIS established what's today known just as HACCP (Hazard Analysis and Critical Control Point)[106] protocols. The purpose of the HACCP system is "to reduce the occurrence and numbers of pathogenic microorganisms on meat and poultry products, reduce the incidence of foodborne illness associated with the consumption of those products, and provide a new framework for modernization of the current system of meat and poultry inspection". The HACCP system was set up for large slaughter facilities with staff to deal with all requirements, of which there are many. For small processors or mobile slaughter units, following the protocol is much more difficult.

An organisation explicitly founded to help small- and medium-sized meat processors (100 employees or fewer) is NMPAN (Niche Meat Processing Assistance Network). It is part of the Oregon State University (OSU) Extension Service and aims to help small ranchers and processors by providing technical assistance, educational resources and information on funding opportunities. Dealing with HACCP is one of the main challenges because:

> "HACCP is not a stand-alone program. Necessary prerequisite programs must be in place before its full implementation. Prerequisite programs, an essential part of the overall food safety plan, are practices and/or conditions needed before and during HACCP. Typical prerequisite programs include Good Manufacturing Practices (GMPs), raw material control programs, vendor certifications, sanitary standard operating procedures (SSOPs), and recall and trace back procedures[107]."

Because "the HACCP system relies on extensive verification and documentation", it is labour intensive and costly, in particular for small and medium size businesses.

Big Meat has found numerous ways to reduce costs. Today, according the US government, "four large meat-packing companies control 85% of the beef market. In poultry, the top four processing firms control 54% of the market. In pork, the top four processing firms control about 70% of the market[108]".

These companies not only control the market through lobbyists, they also hold a lot of political power. If lobbyists can't persuade lawmakers to change the legislation, adding certain exemptions is often an equally effective way to deal with burdensome regulations. In recent years, a number of large facilities have been approved for 'trial programmes' that permit increased line speed, while a new swine slaughter inspection system (NSIS) allows for company employees – rather than government inspectors – to conduct controls. Several organisations such as Food and Water Watch and the Center for Food Safety (CFS) tried to prevent the implementation with a lawsuit which was dismissed by a court in September 2022. In a statement, CFS said: "Despite raising significant dangers to public health from replacing federal inspectors with plant employees (trained and employed by the slaughterhouses), the federal court dismissed the concerns and agreed with USDA that the federal inspection required by Congress would still take place. The industry push for 'self-regulation' goes back decades, and in the 1990s the USDA began moving in that direction. Since the government projects widespread adoption of the NSIS rules (companies producing over 90% percent of the US pork supply), these policies will greatly impact consumers."[109] Or, as the PBS investigative programme Frontline puts it[110]:

"Both inspectors and consumer groups have voiced concern that too much power has been granted to those in the meat industry, that 'the fox is guarding the chicken coop' in other words, and that meat inspectors have been stripped of the power to watch and touch meat on the line. To inspectors critical of HACCP, the acronym has come to stand for 'Have a Cup of Coffee and Pray'".

Chapter 9

MSUs and abattoirs: big and sophisticated or small and simple

How did the US get to this point? Until the 1980s, farms were likely to be diversified, combining arable and livestock farming. Pretty much every small rural town had a slaughter facility and a 'meat locker' – farmers with insufficient freezer space on farm could rent space at the local meat locker to hold their meat until it was needed, for retail or home consumption. During the 1980s farm crisis, thousands of family farms went bankrupt, which of course had an impact on family-run slaughter facilities and processors, too. Add to that the growing numbers of 'big box stores', which were increasingly built on greenfield sites on the outskirts of towns. With easy access to motorways, they soon became a one stop shop for many Americans. Like bakers and greengrocers, many independent butchers in town centres were forced to close.

"In the 1980s, the big meatpacking companies really started to flex their muscles," says Mike Callicrate, who farms just outside St. Francis, a small town in north-western Kansas, just a few miles from the border to neighbouring Colorado.

Callicrate farms about 9,000 hectares. He runs a small feedlot and produces about 800 pigs per year. He says when he started ranching, he had the choice to sell to about 20 different packing companies in Texas, Kansas, Nebraska and Colorado, but by the 1990s only a few conglomerates were left and the cattlemen felt the effect – in the 1970s, 70% of the profit from meat sales went to producers, today, he says, it's

30%. After he joined a class action lawsuit against the food company and meat processor Tyson, with charges of price fixing and acting as a monopoly against the interests of producers and consumers, Callicrate found that he was unable to sell his cattle anywhere.

He wasn't the only one, even ranchers not involved in the lawsuit struggled, says Laura Krebsbach who came up with the idea of a mobile slaughter unit in 2004. In the early 2000s, the trained paralegal was based in Nebraska, working for an advocacy group.

"The big meat processors flooded the market with cheap meat. Farmers either went with these big companies, or they had no access to the markets. The USDA-run facilities closed one by one and for farmers there was no way to get animals to slaughter."

Krebsbach's goal was to give farmers market access, so that they could directly market their meat. A mobile slaughter unit seemed like an obvious solution. To work out what specifications such a unit needed to have in order to be efficient and get the necessary USDA approval, she assembled a small group of experts and farmers – one of them was Mike Callicrate.

HACCP establishes the ground rules: The unit needs to be high enough to hang a beef carcass, the workspace has to progress from dirty to clean, and it has to have a cooling unit. The USDA veterinarian needs to be able to inspect the organs, and a system has to be in place for slaughter waste disposal. The carcass has to be treated with hot water (82°C or 180°F), sprayed with a chlorine wash or a hot water vinegar/lactic acid solution.

Krebsbach managed to source six custom built so-called 'drop deck or drop belly' reefers (refrigerated trailers for articulated lories) from a meat processor who had used them for the transport of sides of beef. The reefers already had lowered axels to provide the required extra height, a reinforced hull and rails on which the carcasses could be moved. A small construction company in Nebraska transformed the reefers into moving abattoirs: a control unit for electricity, temperature, hot water supply and appliances were fitted; a 'dirty to clean' work station was designed; and additional rails for shifting the carcasses into the cooler section were installed. In 2004, the cost of a fully refurbished MSU was $130,000.

The unit can be operational pretty much anywhere, as long as there is access to water and electricity. The infrastructure that has to be in place on site consists of a corral for animals delivered the previous day to rest in overnight, and a chute that serves as a kill box. The MSU can be operated by just two skilled butchers and it can be used for the slaughter of cattle, pigs or sheep.

But setting up an MSU is costly. Mike Callicrate owns two of the six units that were built in 2004. He has built a standard warehouse/docking station for the reefer. Inside the warehouse, rails and chain winches for moving the carcasses have been permanently installed, together with a kill box and a dehairing unit for pigs. In a day, 20 pigs or 10 head of cattle can be slaughtered.

Mike Callicrate works with
a custom built hangar that
has a docking station for
the reefer on one side.

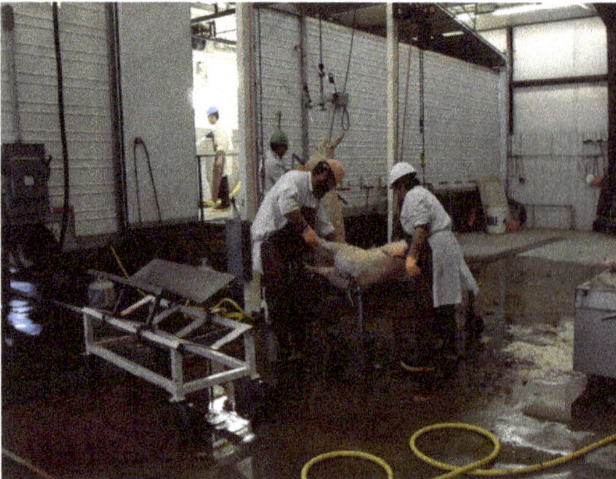

The MSU sits in the hangar, the
animals are bled out outside.

The farm in Kansas is one half of Callicrate's operation. Few people are likely to buy half a pig or a whole side of beef; for consumers, the meat has to be processed, portioned and packed. Curing and smoking adds further value. Because none of this can be done in the MSU, Callicrate built a processing plant and retail outlet 350 kilometres to the west in Colorado Springs. Once the carcasses of the animals slaughtered on any given day are in the cooler, the MSU makes the three-and-a-half-hour trip to the processing plant in Colorado Springs, while the other MSU returns to the farm, ready for use the next day. It is a seamlessly functioning, economically viable system, but it has taken a lot of time and money to put into place. And it shows that an MSU on its own is not sufficient, it is only the first step in the meat processing chain.

Even farmers and ranchers not taking on Big Meat in a class action lawsuit often find it difficult to get their animals slaughtered. In his 2018 book Dirt to Soil, farmer and rancher Gabe Brown writes[111], "The biggest challenge facing most producers who want to get into direct marketing their meat is the processing, and that was true for us, too. When we first started raising grass-finished beef, there were only four slaughter facilities in the entire state of North Dakota that were inspected to allow the retail sale of processed animals. The waiting list to get any livestock processed in these facilities was thirteen months.

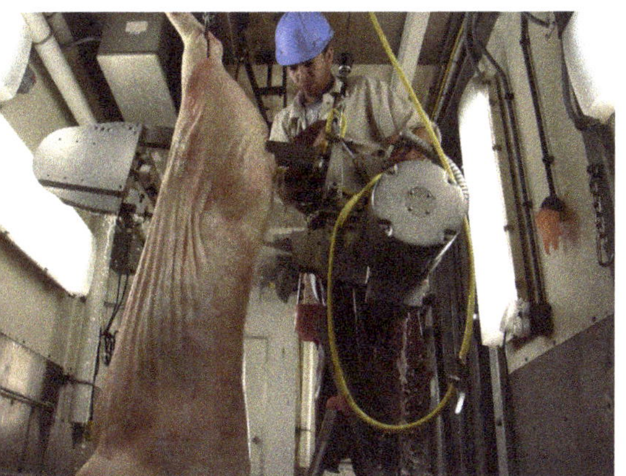

The carcass is split in half.

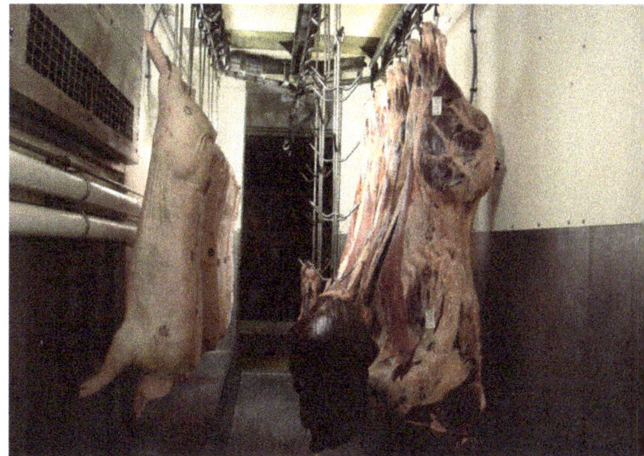

The animal halves and usable organs will be transported to the facility near Colorado Springs for processing.

We realized that we could not operate with those constraints, so in 2012, we joined a group of producers and other investors to form Bowdon Meat Processing (BMP), a cooperative that is inspected by the state department of agriculture to allow for retail sales of meat. It is located in the small town of Bowdon, which is ninety-three miles away from our ranch." The new slaughter facility opened in 2014, says Brown. "We have a standing appointment to have our livestock processed at BMP every two weeks. (...) We transport the animals to Bowdon, and on the same run we bring back home the meat that was processed from the animals delivered two weeks earlier. It works out well for us."

Bowden, North Dakota.
Population 138.

Farmers getting together, forming a co-op to build and run their own slaughter facility – I was intrigued, and in October 2022, we got the chance to visit Bowden (population 138) and talk to board members, staff and butchers.

A sign, reading 'Bowden – Established in 1899' at the village entrance is seasonally decorated with an array of pumpkins and corn stalks. Another two hundred yards down main street, we find a coffee shop and right opposite, next to a grocery store, Bowdon Meat Processing. As we get out of the car, Bob Martin walks over to greet us. He is one of the BMP founders and a board member. We've come on a Thursday, one of the two slaughter days. The farmer who is scheduled to bring his pigs for slaughter this morning has not yet arrived, which means there is time for an unhurried tour of the abattoir. Next to the small office is the processing room for cutting and wrapping, making ground beef for hamburger patties and sausages. To the left is a large cooler for aging sides of beef. The door at the back of the room leads to the kill floor. Behind it are separate holding pens for pigs and cattle – the latter has high, solid metal panels, because sometimes bison are brought here for slaughter too. The pens connect to a passage through which the animals enter the chute which opens towards the kill floor. On the other side of the building are more coolers, the smoker and office space including a cubicle for the vet and utility rooms. One of these rooms is used to store the bins with slaughter waste. Today however, some 20 drums filled with offal and cow heads sit on the roadside next to the facility: slaughter waste is picked up every Thursday by a disposal company from Fergus Falls in Minnesota, 360 kilometres east of Bowdon.

BMP co-founder and chairman Cory Hart.

The pigs still haven't arrived, so we return to the office where we meet Corry Hart, another founding member of the BMP and now also its chairman. Bowdon always had a meat processing facility, Hart tells us. In the 1970s, the owner also bought the grocery store and expanded the slaughter capacity. The business changed hands once more, but for several decades, things went really well. Farmers and ranchers in a hundred mile radius around Bowdon relied on the abattoir which slaughtered several hundred animals a year.

In 2008, the owner/butcher suddenly died of a heart attack, leaving a wife and two school aged children. The Bowdon farmer co-op bought the store and meat plant as an "emergency measure". "We are a small community and we wanted to help the family," says Hart. "But we soon realised that the plant was at a point where it either needed major investment or it would have to close down."

Rather than refurbishing the old facility, they built a new plant with a better layout and more cooler space. In 2012, a group of farmers formed the Bowden Meat Processing co-op, which today has around 70 members, and started raising money. They sold $5,000 shares and applied for state funding and grant money. It took two years to raise over $1.2 million, get the plans approved, source and buy all necessary equipment and get the facility built. "It was an incredible amount of work," says Hart, "but we did it for the community and because we wanted to keep this facility for the many farmers who want to slaughter just a few animals for their family, for their friends and maybe a few local customers."

The first animals were slaughtered in December of 2014. "We need to slaughter 10 heads of cattle a week just to pay salaries," says Hart. At present, they slaughter on average 500 animals annually; 90% are beef cattle, but in autumn farmers bring pigs for slaughter. Occasionally, the slaughtermen have to deal with sheep, goats and bison, too. The facility serves 60-100 farmers who do not have to be members of the co-op – some bring one animal a year, others come very regularly. The plant is inspected by the state of North Dakota rather than the USDA. "The state inspection standard is actually stricter than the USDA standard," says Hart, "but state inspected meat can only be sold within that state, while USDA inspected meat can be distributed nationally." For the farmers the difference matters little: their customers usually pick up the meat at the farm or at the Bowdon facility, and once it is in their possession, they can take it home even if they happen to live in another state.

Bowdon lies at the halfway point between the eastern part of North Dakota with very fertile arable land, in particular in the Red River Valley, and the more arid ranch land in the western part of the state. Corry Hart farms near Bowdon, together with his sons. They have a 470 cow/calf herd. He has about 100 animals slaughtered at BMP, the rest he sells to one of the 'Big Four' meat processors, Tyson, JBS, Cargill and Marfrig. "Most farmers in our region have some cattle and they grow five or six different types of crops, durum wheat, barley for dog food and beer, flax, sunflowers, corn and soy." Another farmer, Mike Flick, joins the discussion. He farms 13,000 acres of arable land and 5,000 acres of permanent pasture near Bowdon, together with his two brothers. The farm has a small finishing yard, but most of the animals are sent to Nebraska for finishing.

Do we want to see the farm? We get into Flick's pickup for a quick tour. As he points out the grain silos and machinery, it becomes obvious that industrial agriculture really is a high stakes' gamble. Climate change is narrowing the windows for sowing and harvesting. To get the work done fast and efficiently, ever bigger machinery is needed. The tractors, combines and haulage trucks we see are worth about $9.5

million, says Flick. He works with a financial consultant to get the best price for his grains – which often means storing it on farm and selling once prices are up. The brothers have, therefore, invested in grain silos with a storage capacity of 1.5 million bushels of corn, which amounts to another investment of about $3.5 million. Annually, the family sells roughly 2,000 heads of cattle to Tyson or JBS, only 50 to 60 will be slaughtered in Bowdon for direct marketing.

There is no way the Flicks could direct market all their animals, at least not in North Dakota which, according to the 2021 census, has a population of just 774,948.

For small farm communities such as Bowdon, autumn is the time of year to woo out- of-state customers – it's duck hunting season! North Dakota is part of the Prairie Pothole Region, there are thousands of tiny lakes and wetland areas, which are an ideal resting place for migrating waterfowl. That duck hunting is serious business became obvious the night before, when we checked into a hotel, a half hour drive to the east of Bowdon. "ATTN: HUNTERS. IF THERE IS ANY EVIDENCE OF BIRDS BROUGHT INTO THE ROOMS, THERE WILL BE AN ADDITIONAL $200.00 FEE", warned a laminated sign at the reception desk.

Mike Flick's farm is a few minutes drive from Bowden.

From the rack next to the reception, I picked up a flier promoting the annual Bowdon 'Duck Fest' in October which includes an art show, Saturday night entertainment with live music, followed by brunch on Sunday. And, in case any visitor to Bowdon should not have heard about the event, two huge yellow duck foot prints and the words "Duck Fest" have been painted

On the farm the animals have a great life.

onto the hamlet's main road. Everybody we talk to at BMP says how important for meat sales it is that hunters taste the meat from locally raised, grass-fed beef and know where to buy it before they return home. Some of the hunters now pre-order quarters of beef to be ready for pick up on their hunting trip.

BMP has five employees. One of them is Jeff whose speciality is making sausages. He's been at the plant for four years and has learnt sausage making on the job. He is in his fifties and started out as a cabinet builder and spent the 16 years flying a crop duster. His wife is from Bowdon and getting a job at the meat plant allowed the couple to return to the village.

On the kill floor, butchers Jacob and Kyle are still slaughtering pigs. They, too, learnt their trade mostly on the job. Jacob Fortney is from Bowdon. He always wanted to farm, but when he left high school in 2002, there was no place for him on the family farm. He trained and worked as a welder for 10 years, first in Fargo, then in Bismarck. He liked the job, but when the slaughter facility in Bowdon was built, he applied for a job as a butcher

The BPM sausages are very popular.

140

and returned home. The company that manufactured the equipment offered some training, and Fortney did a two-day butchery course at North Dakota State University (NDSU) – 25 people were supposed to pick up relevant skills by assisting in the slaughter and processing of just two pigs. He was supposed to learn how to slaughter beef cattle from the plant manager in Bowdon, but it turned out that the person knew little about slaughter and was soon fired. The next two managers were hired straight from NDSU and had more extensive training. Fortney worked closely with both of them and is now the plant's manager[112].

On his two-day NDSU course, he learnt to stun a pig with electric tongs, but he prefers to use a bolt gun. I later asked several other butchers and a vet about this method. They agreed that the chances of a miss-stun were extremely high which also made it very risky for the butchers. The reason lies in the anatomy of a pig: as it ages, the sinus cavity expands, pushing the brain further back into the skull. As a result, it is extremely difficult to place the bolt gun correctly in order to hit the brain directly

Lauren Monson is the office manager...

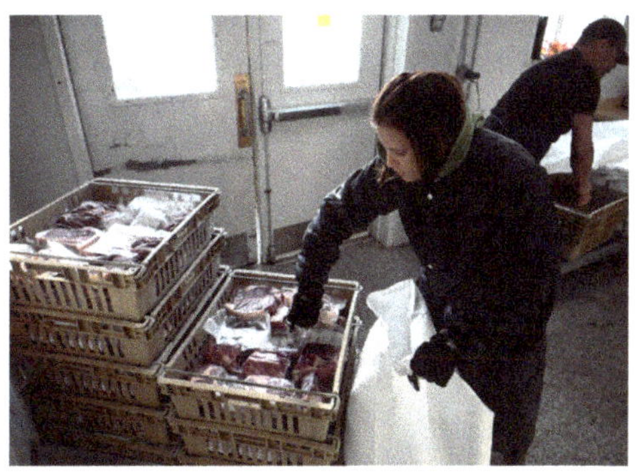

...and helps out wherever she is needed.

A farmer picks up his meat.

rather than just the sinus cavity. It helped us to understand what we witnessed that day in Bowdon, as the butchers slaughtered large hogs. However, the meat inspector was present and did not object.

As one of the butchers has a day off, office manager Lauren Monson has to help out with packing orders. She, too, grew up on a farm near Bowdon and her father brought animals for slaughter to the old facility. "There was a real void when the plant closed," she says. Monson studied to become an elementary school teacher. Ten years ago, she and her husband moved back to Bowden to help out and eventually take over the family farm, a hundred head cow/calf operation. While the butchers are cleaning up, she drives us out to the farm to meet husband Brice and daughters Anna and Elli. The girls are determined to show us the calves they are training for a Cloverbud programme event organised by the agricultural youth organisation 4-H.

After their return to Bowdon, Monson got a job at a nearby school – in North Dakota, a thirty-mile commute one way is still considered 'nearby'. With many young people forced to move to cities like Fargo or Bismarck in search of a job, there were only 10 kids attending the school and together the two teachers had to teach six grades. In January 2022 the authorities finally pulled the plug and closed the school. When Monson heard about a vacancy in the BMP office, she applied for the job on a Friday and started work the following Monday. She is in charge of scheduling – in October slaughter dates are fully booked through to the end of January. With each booking she sends out a form – the cut sheet - for famers to fill out: how do they want the meat processed? Halves, quarters or cut and wrapped? Do they want ground meat, hamburger patties or any of the six different types of sausages that BMP can make? Monson answers the phone, chases payments – and when needed, she helps out packing meat. Does she miss teaching? Not really, says Monson. Teaching during COVID was hard and extremely stressful. Her work now is varied, she has more flexibility and she certainly doesn't miss the daily 60 mile round trip to work, especially not in winter. But most of all it is about what BMP means for the community. The facility is a lifeline for the village because it brings in people – without it there would be no coffee shop, no restaurant and probably no grocery store either.

Marketing the meat remains the biggest problem for the co-op. North Dakota is in drought, which has made managing a diverse farm with row crops and cattle even more challenging. It would really help farmers if they could sell some of their animals to BMP and have them deal with marketing and sales. Co-op chairman Corry Hart is working hard for BMP to become one of the suppliers in the North Dakota 'Farm to School Meat Programme'. According to the government website, BMP could be an ideal partner[113]: "We are hoping more producers will participate in Farm to School procurement opportunities," Agriculture Commissioner Doug Goehring said. "School lunch programmes can help local economies and the state's economy by buying locally produced and processed foods such as beef. Serving North Dakota food products grown and raised by our farmers and ranchers and processed by local businesses, helps students learn where their food comes from." In 2021, four schools bought meat from the Bowdon facility, in 2022, two schools placed a second order. 115 kilos of minced meat may feed a lot of kids, but for a meat plant, it is an extremely small amount. Hart remains hopeful. During COVID, many customers bought meat for the first time directly from farms and small processors and sales are still up. "They want to know where their meat comes from or just want to have meat in the freezer," says Hart. He has attended a number of networking events to increase sales, but he also has a farm to run. Ideally, the co-op would have a dedicated person promoting and selling BMP meat, but so far there isn't the money to employ someone.

In North Dakota, state run initiatives such as the School Lunch Program are not decried as 'socialist' nor is 'co-operative' a dirty word. In this respect North Dakota is rather different from other US states, Sarah Vogel tells me when we meet her two days later in Bismarck. Vogel is a former North Dakota Agriculture Commissioner, an attorney, and she is the author of "The Farmer's Lawyer", which is part memoir, part court room drama. Vogel brought a successful class action lawsuit against the Reagan administration over farm foreclosures.

First stop on our tour through Bismarck is a huge, glass fronted building close to the Missouri River, it's the headquarters of the Bank of North Dakota. It is the only state owned and state-run financial institution in the United States. Its origins go back to the early 20th century, and the Non-Partisan League (NPL), an organisation that aimed to reduce the dependency of farmers on large corporations.

The Farmer's Lawyer, Sarah Vogel, shows us
Bismarck and the banks of the Missouri. Lewis and
Clark spent the winter of 1804 near here.

In her book Vogel tells some of the story of the NPL and its links
to her family. In 1932, North Dakotans voted on an anti-corporation
farming law "which would shape the character of North Dakota for
decades to come. (...) North Dakota voters prohibited corporations
from owning farmland and from engaging in the business of farming
and ranching. This prevented corporations from buying land from
desperate family farmers and gleaning profits from tenant farmers.
However, the law did permit cooperatives composed of 'actual
farmers, residing on farms or depending principally on farming for
their livelihood' to acquire farmland and engage in cooperative
farming and ranching"[114]. North Dakotans may have voted for
Donald Trump twice, but they are still very much ok with farmer
cooperatives, state run programmes and even a state-owned bank.

We leave North Dakota for the Pacific Northwest. While Mike
Callicrate successfully works with an MSU, and Bowden has a
farmer owned meat processing co-op, ranchers and processors in

Washington and Oregon are well on their way to build a successful niche meat processing network.

The Skagit Valley stretches from just north of Seattle to the Canadian border. The moderate maritime climate and fertile soils make the valley one of the most productive agricultural areas in the US: over 90 different crops can be grown here, potatoes and grains, any type of vegetable, soft fruit, apples and flowers – the region is known for the production of tulip, iris and daffodil bulbs. In comparison to the Midwest, farms are small and diversified, most keep some animals, beef cattle, sheep, but also goats. But here, too, farmers find it difficult to get small numbers of animals slaughtered and meat processed for direct sales.

The Island Grown Farmers Cooperative was one of the first to raise funds to buy a mobile slaughter unit. It's been in operation since 2002. In a day, two butchers can slaughter eight to ten head of beef or 40 sheep or 20 pigs. The unit also serves the farmers on the many islands off the coast, which mostly can only be reached by ferry.

Also based in the Skagit Valley is Friesla, a company that builds mobile and small stationary meat processing units[115]. The stationary system consists of individual 'modules' for stunning and bleeding, removing the hide and gutting the animal, processing the meat, freezing and storing it. The modules can be interlocked – a farm or cooperative could start out with two modules for slaughter and gutting the animal, and later add meat processing or a unit for dehairing pigs. On their website Friesla claims: "The systems are USDA compliant and fully integrated with your HACCP plan. Compared to building brick and mortar facilities, our Meat Processing Systems cost up to 75% less and can be relocated, rearranged or added onto as your operations grow. The systems are more cost effective to operate than fixed facilities, requiring significantly less power, water and other resources". The cost: $2.5 to 3 million for a full unit from slaughter to freezer, says Nathan Chase, operations manager at Friesla. The mobile slaughter units are custom fitted into TriVan truck bodies. On the day of our visit, we will be able to see the MSU of the North Cascades Meat Producers Coop in operation, Chase tells us. It will be stationed on a farm close to the Old Highway 99, a few miles south of Burlington. The slaughter team will start work at 8 am.

That morning, thick fog seems to shroud the landscape. Only as we get into the car do we realise that what looks like fog is mostly smoke, which the wind has pushed towards the coast from one the many wildfires still raging in the Cascade Mountains. We are to come to Del Fox's farm, who also has a cut and wrap butchery business in a nearby town. Because the Co-op's MSU is mostly stationed on his farm he has invested in the necessary infrastructure: the yard has been concreted over, wastewater flows into a septic tank that is cleaned on the farm. Farmers deliver animals for slaughter the night before, and Fox has installed separate pens to house up to 11 cattle from three different farms, 20 to 25 pigs, and 25 sheep or goats. The truck is free standing, the electricity needed during slaughter is produced by a generator in the truck and by keeping the truck engine running.

The North Cascades Meat Producers Cooperative's MSU stationed at Del Fox's farm.

The MSU needs water, electricity, the surface has to be concrete or paved.

This morning the team – Del Fox and two of his butchers – will slaughter Angus cattle. From the pen, the first animal is guided through a short walkway with a head restraint at the end. Del Fox shoots the animal with a rifle. The men open the restraint, sling up the hoof so that the animal can be dragged onto the concreted forecourt by a tractor. One of the butchers cuts into the neck of the animal to expose the artery, which is then slit open and the animal starts to bleed out. The whole process takes less than a minute. On this occasion the butchers are being carefully watched by not just one but three USDA staff, the meat inspector who is normally present as well as two vets, who have come for a spot check.

The holding pens have
been custom built.

The head of the animal can be fixated
in the clamp if a stun gun is used.

Butcher Del Fox prefers to
shoot the animals because
they are likely dead and
not just unconscious.

The animal is dragged onto
the forecourt and bled out.

Next step: cleaning.

The carcass is winched
into the MSU.

Once the animal has bled out it is dragged up a ramp into the mobile slaughter unit and the doors are closed. Inside there is very little room to manoeuvre but I'm allowed to squeeze into a corner to observe. The butchers sling up all four hooves of the animal to lift it up and lower it back down on a worktable. Next the head and hooves are removed, then the butchers cut down the middle and begin removing the hide. The chains are reattached to the joints of the hind legs, the carcass is lifted once more and the hide is fully removed. Only then is the sternum cut and the animal is gutted. The meat inspector looks at heart, liver and lungs. Then a small trapdoor near the floor of the van is opened and the slaughter waste is removed from the outside. Once the carcass is empty – except for the kidneys, which stay in – it is split in half with a chainsaw and the spinal cord is removed.

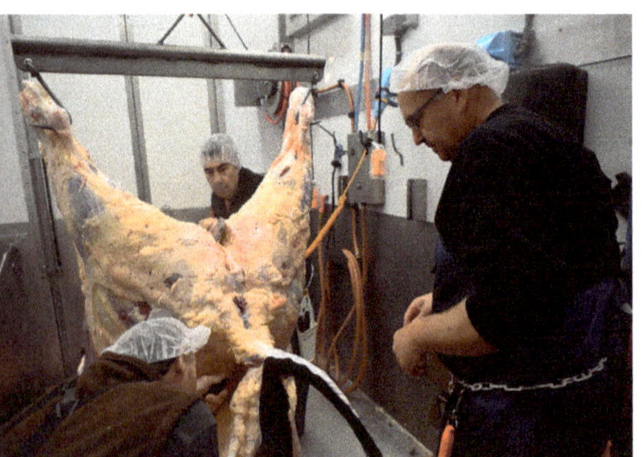

The last of the hide comes off.

The slaughter waste is collected outside.

Before the beef halves are shifted into the cooler at the centre of the MSU, they are sprayed with a 50/50 solution of vinegar and water. The dressed beef halves stay in the cooler for 24 hrs or until they have a temperature between 5° and 7°C (41°-45°F). Later Del Fox will transfer them to a cooler in his butchery where they are aged for two weeks. The farmers then either collect the beef halves, or the butchers cut, process and wrap the meat to the farmer's requirements.

The whole process, from shooting the animal to placing the dressed halves in the cooler has taken three butchers about 35 minutes.

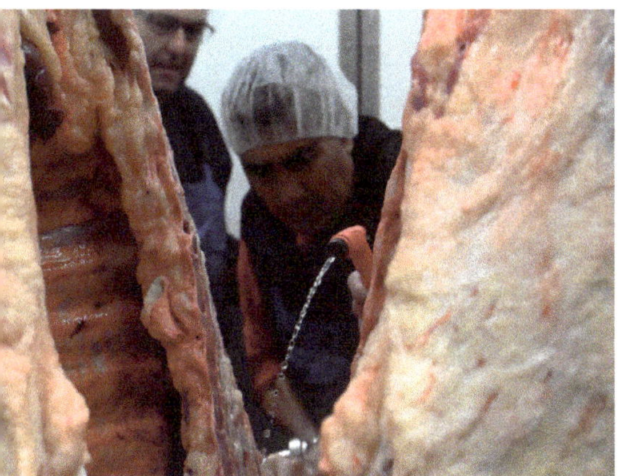

The body cavity is washed.

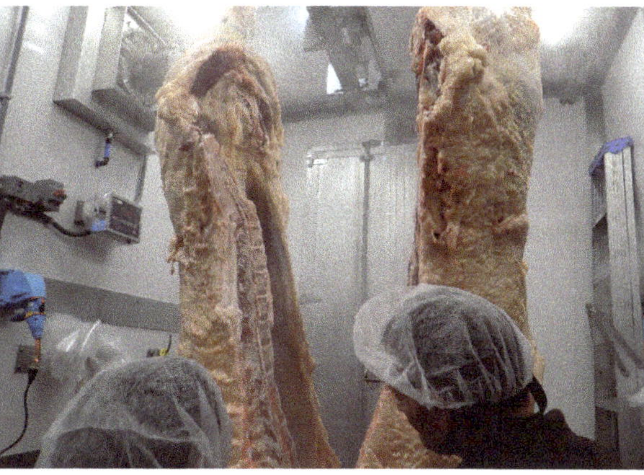

Ready for the cooler.

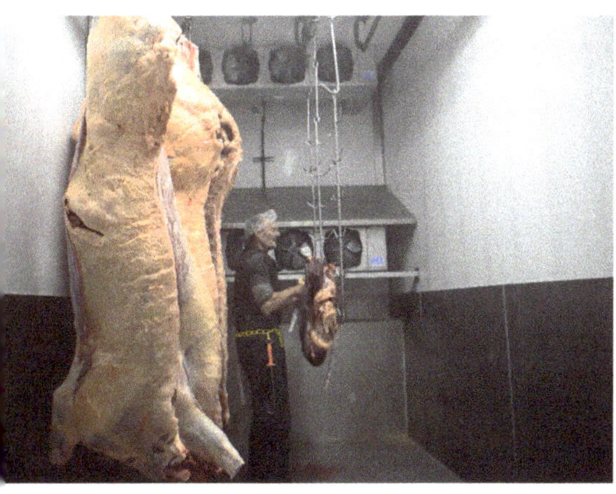

The dressed beef halves stay in the cooler for 24 hrs or until they have a temperature between 5° and 7°C.

A petite woman, wearing a wide brimmed hat and full makeup, is perched on a low wall next to the holding pen. Nancy Hibbing is a volunteer for the North Cascades Meat Producers Cooperative. She spent four hours on the farm the day before to help settle the cattle into the pen for the night, and she will be here until the last animal has been slaughtered. Hibbert shows me the "engine room" of the MSU, housing the control panels, the hot water heater and an old fashioned filing cabinet containing records and the documentation needed for slaughter on the day. The co-op has 65 members, all of them livestock producers. Most deliver between two and 15 animals for slaughter every month, some travelling as far as 100 miles. Non-members can bring animals too, provided there is a gap in the schedule, which usually is not the case. The co-op owns the 'North Cascades Meats' trademark, under which farmers can sell, off farm, on farmers markets or at a co-op run farm stand. But most have their own brands and sales channels.

From next year, Del Fox will reduce the number of days he and his team will be working for the co-op. Butchers and slaughtermen are in very high demand across the Pacific Northwest. Del Fox has been in the business for 39 years. He started helping out on slaughter days at a neighbouring farm when he was in 6th grade. His family was poor, he says, he had to get a job. Following the family tradition of working in construction was not an option – Fox is afraid of heights. He never had formal butcher training but learnt on the job. Nowadays, most of the slaughter he does is on farm and custom exempt, the USDA inspected slaughter in the co-op's MSU is an exception. Fox charges a slaughter fee of $220 per animal, the 'cut and wrap' services range from ¢84 to $1.25 a pound (450g) for beef patties. The slaughter team has a waiting list for 400 to 500 animals, and to expand his business Fox has just invested $325,000 in his own mobile unit. And that means he will be doing even less work for the co-op.

Later that day I talk to the co-op's president, Luke Conyac over the phone. When the co-op was founded in 2013, the main aim was to increase the slaughter capacity, he says. To build the MSU took a year but "that was the least of it", raising funds, getting non-profit status and USDA approval for the workflow from slaughter to cooler were much more challenging and time consuming.

The coop has leased some land and the use of existing buildings on a local farm. "An MSU only makes sense when it is stationary at a facility that can receive animals from other farms or at a farm that has a lot of animals to slaughter in one go," says Conyac. The mobile unit will be moved to the leased land in the coming weeks. Some of the infrastructure has still to be put into place but he hopes the MSU will be operational from January 2023. His biggest headache at present is finding trained slaughtermen to operate it. The co-op has three full time employees and 12 active volunteers. Conyac says he can slaughter hogs, and he hopes he will be able to train some of the volunteers to help. For beef cattle, he is working on an interim solution. "We are in discussion with a third party for slaughter and cut and wrap," he says.

The "third party" turns out to be the brand new cut and wrap facility of the Island Grown Farmers Cooperative in Burlington, which we are scheduled to visit that very afternoon.

Chapter 10

Processing meat efficiently and building a local food system

For historical reasons, in the US, a distinction is made between slaughter, cut and wrap and value adding. When farmers still mostly slaughtered on farm, it was the job of the butcher to cut the meat into steaks, roasts and ribs, mince the rest and store it in a 'meat locker'. Smoking hams, curing bacon or making charcuterie was (and to a degree still is) the job of specialised 'value adding' facilities.

The Island Grown Farmers brand new cut and wrap facility is situated in a business park on the outskirts of Burlington[116]. It was built in 2021 and Travis Stockstill is the operations manager, in charge of the Co-op's MSU as well as the cut and wrap plant. The MSU is operating four or five days a week and has a rotating crew of three butchers. Out of the 16 staff not all are qualified to work in a mobile facility, because apart from knowing how to slaughter, the butchers also have to have a commercial driver's license that allows them to drive a trailer rig. The MSU operates almost exclusively on the islands and "nobody should do that job day in, day out, people need to have a work life balance," says Stockstill. "The days in the MSU are really long. The USDA inspector gives us eight hours of slaughter time per day, and you need to add to that an hour for setting up in the morning, and at least an hour for cleaning in the evening."

The Island Grown logo is prominently displayed.

Before the Co-op agrees to send out the MSU, a team will inspect the farm: is there a holding facility for the animals; has the unit access to electricity and water; or will they have to work with the on-board generator and their supply of potable water? These days, the unit visits mostly well-established host farms on the islands. Neighbouring farmers deliver their animals the night before or call ahead and let the team know their time of arrival. The unit has cooling space for 12 to 15 heads of beef. Annually, 2,500 animals, beef, lambs and hogs are slaughtered in the MSU.

The Island Grown Farmers Co-op has 80 members, 60 of whom are using the MSU. As with the North Cascades Meat Producers Cooperative, the main issue is the lack of slaughter capacity.

Stockstill has requests for an additional 75 slaughter days from within the co-op, and at present 21 farmers are on a waiting list hoping to join. Demand varies, some farmers would like to bring 10 heads a week for slaughter, others just one or two occasionally.

Direct sales through the Co-op are limited to the meat stored in the fridges and freezer in the small waiting area next to the reception desk at the processing facility in Burlington. All farmer-members have their own sale avenues into restaurants, schools and farm shops, as well as direct sales and farmers markets.

Before we can enter the butchery section, Stockstill kits us out with hairnets and white coats. The six butchers working on the floor are getting close to the end of their workday after cutting and processing six lambs, 11 pigs and 300 pounds (136 kilos) of trim for three different farms.

David Stockstill's office is small and cramped. The monitor allows quick communication with the shop floor.

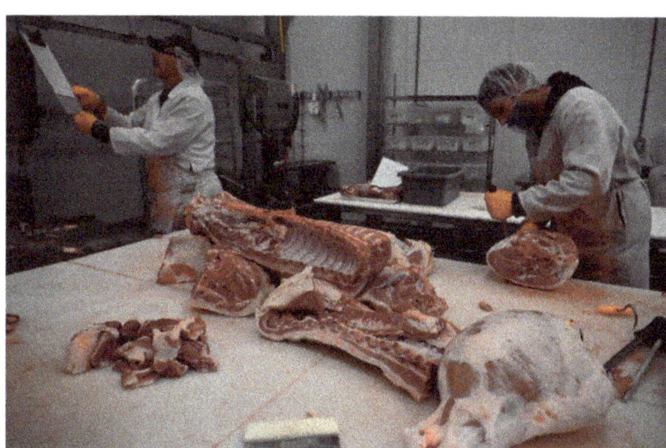

The butchers have processed six lambs, 11 pigs and 300 pounds (136 kilos) of trim for three different farms today.

Traceability is maintained throughout, from slaughter to packaged product: each farmer will receive the meat of their animals. "It's great if the farmers bring a few animals at a time, but if they bring one pig, we work with that, too," says Stockstill. Farmers fill out cut sheets specifying the cuts they require and what should be made from the trim, which can be made into ground meat, sausages or sent on to a processor who makes value added products such as salami. Stockstill works with Cured Processing in Wenatchee. Director of operations there is David Zarling who also is the NMPAN program director as well as a central figure for the meat processing network in the Pacific Northwest.

Processing costs vary: farmers pay a slaughter fee, a fee for cutting, boxing and an extra fee if they require 'specialty' cuts such as oyster steaks or Vegas strips.

Once the butchers have processed the meat, it is frozen. Farmers have two weeks to pick up the meat, which is either ready packed in cardboard boxes or left in crates for the owners to sort – which can be helpful if specific orders for restaurants have to be fulfilled. Anyone failing to pick up their produce in time will be charged for every additional storage day.

As the butchers start cleaning up, we return to the office. Now aged 39, Travis Stockstill has worked in MSUs for several years and has extensive experience processing meat and producing charcuterie. I ask him whether he got into the meat industry by choice or by accident. "I come from Santa Barbara," he says, "I was a surfer dude and a vegetarian from age 16 to 22, I didn't believe I'd ever work in meat." He became a journeyman tiler and had plenty of jobs

Equipment needs not just to be clean but also easy to access.

until the financial crisis hit in 2008. "Nobody was building anything," he said. At the time, Stockstill's brother was the general manager of a slaughter facility in Vermont. "He called me and said, 'Come out here, you get a new career,'" says Stockstill. It took him two years until he felt like a skilled cutter and slaughterman. Only after the company invested in his HACCP training and humane handling certification did he finally feel that he could have a career in meat. He returned to Los Angeles where he managed a boutique butcher shop and met his wife, who is from the Pacific Northwest. When the couple decided to start a family, they moved back to Washington. Stockstill started work on a USDA certified mobile slaughter unit and ran a smokehouse for a few years, until he heard about the new Island Grown Farmers Co-op facility and he applied for the job as manager. Food safety matters to him a great deal, and he likes the job security. "People will always want meat," he says. But wages are a big issue in the industry. "In the 1980s, a meat cutter would have been able to put his kids through college, have full benefits and a pension," he says, "now we expect people to ruin their health for a minimum wage." At the Island Grown Farmers Co-op everyone is paid at least a living wage. "That's why we have great retention of staff and that's why we can invest in training people."

Wages and training are two of the topics Travis Stockstill and fellow butcher and friend David Zarling frequently discuss on their 'Meat Block' podcast[117].

Zarling's career as a butcher and in meat processing began with his passion for food. The now 38-year-old grew up in Michigan. "In our family, we are all tall, we are big, you clean your plate and you get another plate and you clean all that, too, that's the mentality. I love food. I love well done food, I love the culinary world," says Zarling. The expectation was that he would "go to college and have a fancy job". His family moved to Chicago when he was a teenager, and Zarling did attend college, but not for long. "In my 20s, I realised all I want to do is work with my hands and I want to get into food. I was not in a financial position to go to culinary school. So, I got a job as a dishwasher. I knew I wanted to work with meat, I was really influenced by a fellow called Rob Levett in Chicago, at a place called Butcher & Larder. At the time, Rob Levett's concept was totally new: snout to tail butchery. Whole carcass utilisation was an extremely attractive message to me. It was the relationship with Rob, walking into the market, having this bacon that he made there, from this farm, this style of hog, and he could say, try this 'jamon evirgo' and try this thing that I made and tell me what you think. It was the deconstruction of a raw material and then its reconstruction into something beautiful that I loved. And there was the interaction with customers, it was the community - I just loved all of it."

Zarling also got interested in sustainable and regenerative agriculture, in foraging and 'hyper-local' food. He finally got a job in a commissary kitchen. "All I got to do was cutting portion steaks for chefs and cutting briskets for the smokehouse and cutting the pork butts for the al pastor (spit grill). It was a job nobody wanted and I took it and liked it. I got to work in a cold room by myself, I didn't have to talk to anybody, I got to listen to my own music, I loved it. I was very starry-eyed about the industry, I wanted to directly work with farmers, bringing their produce to market and I thought craft butchery was the pinnacle of the industry." He decided to leave Chicago for the Pacific Northwest, with its 'foody' culture and high-end restaurants. His first job was as a kitchen portion meat cutter for a high-volume retail chain: "I learnt how to supervise, I learnt how to add value and how to market to people, to create things for people to take home and cook. Out here in Seattle, in the tech industry, everybody just wants to take food home, put it in the microwave, take it out and eat it". One of the restaurants Zarling worked for was Agrodolce, at the time owned by James Beard award-winning chef Marina Hines, who also opened Tilth, the first certified organic restaurant in Seattle.

"We sourced everything locally, from the Pacific Northwest, and meat was directly purchased from farmers. We got Skagit River Ranch Beef, which today is processed by Travis, and from Jeff Rogers. One day I got a call from my old chef, saying, hey, our lamb producer built this slaughter truck and he's looking for help. Somebody who can throw the guts out. So, I joined up with Jeff who's become a dear friend of mine and I learnt how to slaughter on his truck."

With his love of food and farming, his work ethic and his boundless energy and curiosity, Zarling had not only acquired skills as a slaughterman, meat cutter and chef, he had also built a network of like-minded people, both in agriculture and the food and restaurant industry. One of them was Travis Stockstill with whom he worked on a mobile slaughter unit in Bellingham for a while until Stockstill went to another job. Zarling moved back to Michigan to run a farm together with his partner, as well as manage a small slaughterhouse. "I was there for a few years, managing a small, family-owned slaughterhouse, doing everything under that roof. Raising my sheep with my partner: she did the vegetables, I raised the animals." In 2015, he was headhunted by a meat company in California. Animals were raised on the company's own farm, butchered and processed in their own 20,000 sq. ft. processing facility and served or sold in one of their seven restaurants and butcher shops. "I wasn't even looking for a job and they tracked me down and said: hey, we heard what you did at this family farm in Michigan, we think you might be a good fit for our company. Could we fly you out 1st class, wine and dine you and tell you about our operation?" Zarling took the job, but wasn't happy with it for long. "They were scaling to a level that I was not comfortable with professionally, it didn't align with my values any longer, and I wanted to get back into that niche space".

The niche meat sector is where he has found his place, running a business, teaching and consulting.

Wenatchee is a town roughly 150 kilometres east of Seattle beyond the Cascade Mountains. Driving there needed some planning, as the Bolt Creek fire which started in early September was still raging and depending on the direction of the blaze, roads were being closed at short notice, mostly just for a few hours but sometimes for several days. Wenatchee is situated in a fertile valley on the bank of the Columbia River and was once known as 'the apple capital of the world', a title now contested by the town of Yakima, a further 100 kilometres to the east.

We talk with David Zarling in his office at Cured Processing, which he has been running since 2021. The business is owned by the Visconti family in nearby Leavenworth – a tourist town and ski resort that tries hard for an Alpine flair, down to the Bavarian style houses. The Italian family restaurant Visconti's has long been known for its excellent charcuterie. Initially, the family made sausages and cured hams in a basement kitchen under the restaurant, but the charcuterie was so popular that in 2020, they decided to build a dedicated meat processing plant in Wenatchee.

"We are a value-added processor or co-manufacturer, we are trying to extract the most value possible out of the chain," says Zarling as he walks us through the production. This morning, some of the staff are making sausages which will later fill the three smokers: two hold 400 pounds (about 180 kilos), the third has an 800 pound capacity. Next door, we meet Victor, who worked at the restaurant in Leavenworth for nine years and now creates all the spice mixes. His colleague Hector is in charge of maintenance – not an easy task in a facility where temperature and humidity have to be tightly controlled. There are different spaces for fermented, dry cured and semi dry products to mature and for curing and smoking bacon. Every product can be portioned, wrapped, packed and individually labelled.

Processing is only efficient if smokers and cookers are filled to capacity.

David Zarling doesn't care about the looks when HACCP rules are involved.

The 25 staff have very different backgrounds. Some have worked in the Visconti restaurant kitchen, but most were previously employed in the apple orchards and lost their jobs when much of the apple industry shifted to Yakima. "They are very hard working and have a lot of transferable skills," says Zarling, who nevertheless trained everyone from the ground up. Central to everyone's work is a large screen in the hallway. Transparency is key for customers and workers, says Zarling, which is why all available information is accessible on digital spreadsheets: from customer information to time and date of meat deliveries. At a glance, anyone can see on screen where in the process a particular batch of meat is and where it will go next, storage space can be allocated and storage levels checked. Every worker has been trained in how to access and use this information and that gives them responsibility, as well as control and independence. Eight of the 25 staff now lead their department, says Zarling. Everyone is paid at least a living wage, which in Wenatchee is $18/hour. There are incentive bonuses for accuracy, and Zarling hopes that in a few months every position will come with full benefits, too. A development programme establishes what training an employee might need to take on a different and/or more responsible role. If it were up to Zarling, the company eventually would be worker owned.

In the US, Cured Processing is the one and only pilot for a small-scale co-packing plant in which as little as 400 pounds of meat can be processed cost efficiently. A volume of 1,200 pounds a day would be optimal, because it would fill a smoker. Normally, the minimum for a co-packer is 5,000-10,000 pounds (2,250-4,500kg) of raw materials. "To be profitable we need to aggregate," says Zarling, who is in favour of more automation – not to replace butchers but to make the physical work easier. "I believe that there is a way to bring the wisdom from lean manufacturing to small production without making it a soulless machine." That means: do not compete on price. "The American model is all about cutting costs and getting cheaper and cheaper and that's why American manufacturing isn't worth a damn anymore. But if you focus on the culture of the business and the process, you will figure out how to get to a net profit" For Zarling, there is no 'right size' for a business, everything depends on doing the math. "You can have a profitable model with five people in a crew and a hundred people in a crew – and there are unprofitable versions of both. You have to figure out: I need to make this much product at this price, this often with this many people, to have this bottom line". At Cured Processing, the sums work because the company has a variety of customers who are very different in their demands and the volumes they deliver. "We've got three or four customers who are start-up companies who have significant volume, if they send me an order it's going to be 2,000 to 6,000 pounds at a time. That's what pays my bills. That allows me to take on somebody like Casey, from BCS livestock just north of here, and make a small batch of jerky. I am going to lose margin on that, but I can provide that service because I've got this other steady volume."

Cured Processing has an own brand, but most of the work is co-packing – producing for another brand. Customers can be farmers who send their meat to Wenatchee for processing – we saw an example at the Island Grown Farmer Co-op where Travis Stillstock regularly ships some trim to Wenatchee because farmers require value added products that cannot be made in Bellingham. Other farmers are close to Wenatchee: "There is a regenerative farm near here, they raise non-GMO pastured hogs on a hazelnut finish program, we do small batches of very high value culatello hams for them." Zarling can afford to do that because there are companies who put in regular orders for a product with their label – for example, a brewing company that sells a particular type of sausage to go with their range of beers. To make them, Cured Processing orders premium pork from Iowa "at a good price". Cured works mostly with customers based in the Pacific Northwest, but because they are so highly specialised, they also work with companies in other states.

A sophisticated labelling machine is expensive...

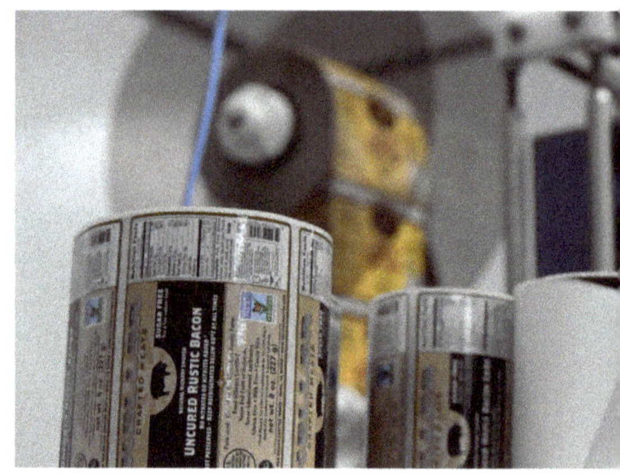

...but essential to deliver shelf ready products.

In future they will no longer deliver their own brand product – there are distributors for that who do it more efficiently and cheaply, says Zarling. They've also invested in packaging - their products have a shelf life of 120 days without needing to be cooled.

"My background really is in person-to-person small scale slaughter and butchery. That's where I came from. And being in that industry as long as I have, I have seen so many facilities close their doors and fail because they don't understand the finance behind it. That's why I am piloting this small manufacturing model, because I think even the smallest producers can do it like this – it takes a bit of the sexiness and the romance out of slaughtering the beef in the middle of the field and selling it to people, but if everybody can cooperate on a level of efficiency, I think people can stay true to their values and deliver a product that is going to be adding to the whole value chain."

Zarling has worked along the whole meat chain, from raising animals and slaughter, cut and wrap and charcuterie to retail butchery, commercial kitchens and high-end restaurants. During his time working as consultant with Michele Pfannenstiel[118] who specialises in food safety and HACCP training, he facilitated company audits and helped to develop a plant management training. It's this wealth of experience he now brings to his 'other' job as programme director at NMPAN.

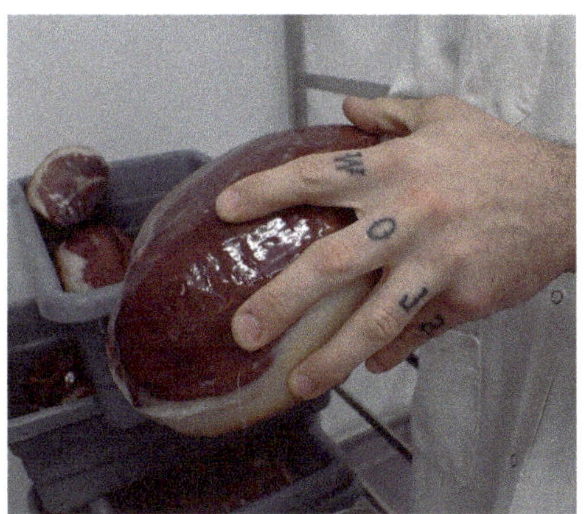

A Cured culatello ham.

Sampling the produce at the Pybus Public Market in Wenatchee.

A business can only work if and when the business model "shakes out on paper". Of course, you can muscle through, says Zarling, "you can make it work, as a business owner, you can work 60, 80 hours a week, but you can't expect the same of a team". He prefers to work with existing businesses: "I think that a lot of the new business owners that I work with have a very romantic idea about what this is. I prefer to work with the existing facilities, who know the hardship and know where the pain points are and then we can say: look, if you try this different thing, I think we can turn this around and expand your margins. Because a lot of times it's just a question of culture and financial hygiene. If you have an operating budget that truly reflects what you really do every day, then you can make decisions – is this viable or not and what do we have to invest to make it work? Being able to have that discussion and plan ahead, I think that's scary and people don't want to talk about it. We do a lot of band aid solutions in this industry, but we don't do a lot of projections and forecasts."

NMPAN together with the American Association of Meat Processors (AAMP), and Kitchen Table Consultants (KTC) have developed an eight-week course which will be launched in spring of 2023. "The Meat Processor Academy" is a boot camp for small niche meat processors to get an introduction into best business practices," says Zarling, who will be teaching several of the courses. The curriculum ranges from business planning to financial management, human resources and marketing and distribution[119]. A similar programme (but aimed at farm businesses) was launched in October 2022. Zarling is particularly excited about the Meat Processors Academy Mastermind Program[120] which will start in February 2023. It will be a 12 months course with no more than 15 to 20 processors. Each will be paired up with a business coach from Kitchen Table Consultants (KTC) with the goal to have an 'action plan' by the end of the year. There will be regular online meetings with peers and teaching modules covering topics from creating a pro forma operating budget to staff training, accounting and tax law.

NMPAN already has a huge back catalogue of webinars, videos and fact sheets. It also provides advice to people who want to apply for grant money[121]. In January 2022, the Biden administration allocated $1 billion to expand the independent meat processing industry in an attempt to break the stranglehold of the big meat conglomerates. "For a small operation, it is very hard to access such funds," says Zarling, "but if you have a training plan and can show how the money is used, you do stand a better chance." But he is sceptical that the huge sums that are at play at present, will change things in the long term. "A lot of grant money from the USDA is funnelled down to the meat processing space to help fortify it and create new infrastructure based on the response to the pandemic. It's money for buildings and now work force development. This is a cycle we have seen re-enact itself in 30 to 40 year spirals. It goes like this: we don't have enough processing capacity, let's put a bunch of money into facilities, equipment and training. Three to five years later we have a bunch of empty buildings that look really nice and a bunch of used equipment on the sales market because the businesses weren't profitable."

NMPAN is also a great place for networking and Zarling suggests we visit two successful small meat processors who have made their businesses work in very different ways: The Meating Place and Revel Meat Co., both near Portland, Oregon. But before we head south, we have to meet his friend, the entrepreneur, sheep breeder and mobile slaughter unit developer Jeff Rogers.

Redmond is on the eastern outskirts of the urban sprawl that Seattle has become. We meet Rogers at the gravel lot next to one of his pastures. He is getting his mobile slaughter unit ready because he will start harvesting his lambs the following week. Rogers didn't start out as a sheep breeder. As a kid, he was a math geek, interested in science and animals. After studying at various universities, not quite sure what he wanted to do with his life, he founded a financial service software company. By the year 2000, the company had offices in California, New York City and Zurich, but Rogers decided this was not the life he wanted and sold the business. He and his wife moved to the Utah mountains, bought a sheep dog and finally some sheep – a mix of Rambouillet-Merino and Columbia. The sheep spent summers in the mountains and were moved to pastures in the plains for the winter. "The sheep were a commodity breed," says Rogers, "there was no money to be made with them, the ranchers were just price takers." But sheep breeding had become a real passion and Rogers decided to go for a breed that would thrive in a humid and wet climate like that in the Pacific Northwest. The couple upped sticks once more and researched sheep breeds in the UK. Today, they are the only breeders in the US raising mules, crossed from Blue Face Leicester and Clun Forest sheep. For the last 14 years, the farm has been certified organic. "I like a challenge," says Rogers with a grin, "raising organic sheep is really difficult." Unlike in the UK, where under organic rules an animal can be treated with antibiotics if necessary and, once the required wait time has passed, will still be considered organic, in the US such an animal would automatically lose the organic status for good.

Rogers harvests around 2,000 pounds (900 kilos) of wool every year. Because the owner of a small wool shop in Redmond is keen to stock local organic wool, Rogers sends the fleece across the country to Vermont for scouring, to one of very few US wool mills that process organic wool. But the mules' real value is in the meat. Initially he was able to have the lambs slaughtered at an abattoir just two miles down the road. Then the ownership changed and Rogers had to find a new slaughter facility. For three years he drove the lambs four hours north to be slaughtered in the Island Grown Farmers Coop MSU in the Skagit Valley. He hated having to transport the lambs over such a long distance and he wasn't the only one struggling to find slots for slaughter – neighbouring farmers were facing the same problem. But discussions of buying their own MSU got nowhere, everybody thought it was a brilliant idea, but nobody wanted to put up any money, says Rogers. In 2012, he decided to go it alone and design and build his own MSU tailor made for the slaughter of sheep. In autumn of 2013, he was able to slaughter the first lambs right next to their pasture.

The unit looks nothing like the huge MSU we've seen in the Skagit Valley – it is much smaller and consists of two parts, a harvest and a cooler unit. Rogers used two cargo trailers: one houses the cooler unit, two AC units and a generator, the second one serves as harvest unit and is fitted with a tank for potable water, a water heater and a pump. For slaughter, the harvest and the cooler unit are linked, but the cooler can be decoupled and hooked on to a pick-up or SUV for meat deliveries.

Jeff Roger's two part MSU for sheep.

The lambs come through a run on the side of the unit.

The MSU is fitted with two AC units, a generator, a tank for potable water, a water heater and a pump.

The cooler unit is detachable and can be pulled with a station waggon.

Rogers talks us through the process. He slaughters the lambs when they are about seven months old, usually seven or eight animals in one day. On a few occasions, he had to slaughter 17 or 18 to fulfil an order but he thinks this is too much and there is a danger of things being rushed. The animals are brought to a small enclosure. Then one animal is led through a short walkway made from wooden panels along the side of the harvesting unit, it is unable to see what's going on. When the lamb reaches the end, Rogers stuns it and uses a wheelbarrow to move it around the corner to the entrance of the unit, slings up the hind legs and hauls it up. Standing on an elevated grid he severs the artery for the lamb to bleed out. The blood is caught in an aluminium basin underneath. There isn't much blood, says Rogers. What there is will be collected in a bucket and later composted.

Inside the abattoir section.

The hooks for the carcass move easily on wheels normally used for inline skating.

Inside the unit, the carcass can be moved from station to station on a steel pole, removing hide and guts, cleaning the inside with hot water and finally splitting the carcass in half. Rogers has designed a vertical bar with hooks at both ends where the hind legs can be fixed. At the top, he attaches wheels which are normally used for inline skates. In his unit, they make it possible to move a carcass easily and noiselessly. On the back of the left side of the trailer is a door that connects the harvest unit with the cooler into which the dressed halves are pushed. There is enough space for 24 lambs. They will be aged for four days at 32°F (0°C) before he delivers them to restaurants and PCC community markets, a natural and organic grocery chain in the Seattle area. Rogers slaughters 200 lambs a year, but the demand is much higher. He would like to increase the flock size, but he has only got 80 acres of land and so far, he has been unable to lease or buy more.

It cost $90,000 to build the units in 2012, today it would probably be twice that, says Rogers. On slaughter days, he has two people to help him inside the harvest unit, but only he himself stuns and kills the animals. Everything has to be really quiet and calm, he says. I think of Mechthild Knösel and her farm near Lake Constance. Does Rogers talk to the lambs? Yes, he does, and he thanks them. "For me it's about respect for the animals, there has to be dignity to their end."

We leave Jeff Roger's farm and make our way to Interstate 5 South. Our next stop will be the The Meating Place in Hillsboro, about 30 kilometres west of Portland, Oregon. David Zarling has put me in touch with Marena Gray, the manager of the custom butcher shop. He suggested a visit because The Meating Place is about to become a one-stop-shop for farmers and consumers: it provides a custom slaughter service, meat processing, a retail outlet and there even is a restaurant.

A few steps lead onto a porch that runs along the length of the building. The faux wooden front and the big 'Meating Place' shop sign add a hint of western charm and cowboy flair. The retail space is huge. With a wooden floor, a dark wood ceiling and wall panels, it keeps to a western vibe, but this is contrasted by the clean, modern look of metal airducts and pipes, and the long brightly lit cooling counters that run along the walls. On display are fresh cuts of beef, pork, lamb, as well as poultry. There is a section with marinated items for barbecuing and ready to eat items such as cold cuts

Pick your own.

The Meating Place west of Portland, Oregon.

and sausages. Even with more coolers and display stands in the centre, the space does not feel crowded. Before I have time to take it all in, Marena Gray is there to meet us. We enter the butchery through a door at the back of the shop. Two of the butchers are cutting down animal halves, others are preparing roasts and steaks, while the trim is ground. Marena Gray is not the only female butcher. She introduces us to Ellen who is working full time as a cutter.

Starting out in kitchen prep Ellen found that she liked working with meat. She learnt meat cutting at a hog processing plant, and three years ago she applied and got a job at the Meating Place. She likes that she is able to work on all stations of the cutting room. The one thing that really irritates her is that all the fat that is

Custom cutting at the Meating Place.

More women are becoming butchers.

No one cuts you slack just because you are a woman.

trimmed off the carcasses goes to waste. She took some of it home and started experimenting in her kitchen, making soaps, hair oils and face creams which she is now selling through her company, Fat Chap.

Portioned meat and mince are stacked in grey plastic trays and moved to the wrapping unit which is managed by Shona. She worked at the Fred Meyer supermarket chain and came to the Meating Place two-and-a-half years ago. It's hunting season – the busiest time of the year, and freezer space is in short supply, but Shona has developed a system of organising and stacking produce that makes it easy to fulfil orders efficiently.

The most important tool for a smooth workflow is the cut sheet: it documents all the necessary information, and because it is handed from department to department along with the meat, each carcass is traceable.

The endless sausage needs to be tied of and hung.

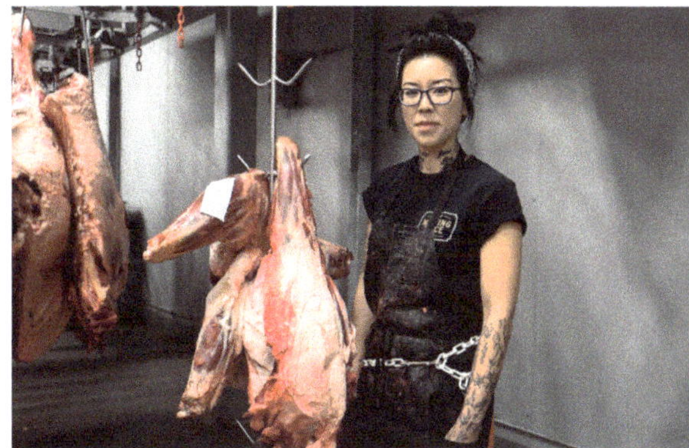

Shop manager Marena Gray standing next to the bear hanging amongst the beef.

Around the corner from the cut and wrap floor is a large room dedicated to sausage making, curing and smoking. The wall it shares with the retail space has a large window. The two butchers filling sausage meat into a casing at record speed have little time to watch what's going on in the shop, but customers waiting to pay and pick up their goods have an opportunity to see butchers at work.

Our next stop is the cooler. It is huge, but this time of year, a lot of space is taken up by game – deer, but also elk and bear. Annually, the butchers process around 830 head of beef, 400 lambs, 300 pigs and 2,500 game animals.

The Meating Place has a long history. It was initially set up as a cut and wrap operation, mostly for game processing, including fish, and staff had a reputation for making jerky, sausages and smoked salmon. It's where Casey Miller, the present owner of the business learnt the craft from a second-generation butcher. In 2011, the Meating Place was reopened in its present location, initially operating only for a few months in the year to process game. Two years later, the retail shop and the cafe were opened which meant there was enough work for the butchers year-round. As the cut and wrap facility, The Meating Place processed carcasses from custom exempt slaughter – meat and meat products had to be returned to the rancher or hunter, but could not be sold[122]. Miller therefore had to buy in USDA certified meat which the butchers then turned into meat products for the shop and the cafe. A butchery does not have to be USDA certified for processing meat for retail, but butchers need to follow HACCP protocol wearing gowns and hairnets while working. Neither is obligatory for custom exempt butchery.

Having their own meat supply has been a long-term goal for Casey Miller and Andy Turner, one of his partners. Turner has a degree in Agricultural and Animal Sciences from Oregon State University, he learnt how to kill and butcher at The Meating Place, and he now runs a farm. At present the team operates a small mobile slaughter unit, but just this year, Turner has invested in a large MSU which will be USDA certified and stationed on the farm. In future, produce in the shop and the cafe will not only come from animals raised on Turner's farm, but the team can also offer a USDA certified slaughter service to ranchers who would like to direct market their meat.

The USDA certified MSU will be the final link in the chain that connects consumers to farmers and ranchers: the animals will be slaughtered in the MSU close to the farm, the meat will then be processed at The Meating Place, and all products will be fully traceable.

Zack is the deputy manager of the shop and really looks forward to the team having more control over meat quality and cutting all ties to large meat suppliers and slaughter facilities. The Meating Place already has an excellent reputation, some customers come from the other side of Portland, a 45-minute drive away. There are some corporate customers who buy meat for their canteens, and every Thursday local sheriffs ride up in a motorcycle convoy to have lunch at the cafe.

The Meating Place employs 60 people, which includes everyone working in processing, sales and the cafe. Everyone is paid at least a living wage and gets healthcare and retirement benefits.

Not everyone who works in processing is a trained butcher. Marena Gray trained as a line cook, worked as a waitress and then in retail. Her first job at the Meating Place was at a counter in the shop. Then she managed the cafe, but she told Casey Miller that she wanted to learn cutting and slaughter. "Everyone here gets a chance to try out different jobs," she says, "we want people to do the work they are best at." Gray got to do trial shifts on the MSU, but she is of slight build and not very tall. "I just can't lift as much as the others and I found it difficult to move equipment, especially when everything got stuck in the mud. The Meating Place is an equal opportunities employer, but no exceptions are made for you just because you are a woman." She tried meat cutting and really liked the work and has now been working as a cutter for five years; for the last three years, she's also been managing the department. She likes hard work and multi-tasking as well as training and people management, she says. And with that she is off to get some work done while we head to the cafe to get a late lunch.

Portland is known as a city of foodies and Providore Fine Foods in northeast Portland is where you might head in search of artisan cheese, groceries imported from Europe, local wine and freshly made pasta. Fresh farm traceable meat and meat products are available from the Revel Meat Co

The signposting definitely helps.

counter. They are produced at the slaughter and meat processing facility in Canby, a small town just south of Portland – and that's where we meet Ben Meyer, one of the owners.

The unit is housed in a non-descript building on the side of a country lane. It's a cold morning and we are happy to wait in the tiny office as Meyer is sorting out issues with a delivery and a boiler that isn't functioning properly – the building is old and there is always something that needs fixing, he says. So, why did he give up running successful restaurants in Portland to manage a slaughter and meat processing facility?

Meyer tells us his story as workers pop in to ask questions, the delayed Whole Foods driver finally shows up and someone else is in search of a band-aid. He is from Fort Wayne, Indiana, a 90-minute drive southeast of Chicago, but he went to school in Minnesota. It was his great-grandma who inspired his love of food. She canned vegetables, cured hams and, being of German origin, made no less than 50 barrels of sauerkraut every year. "My mother really admired my great-grandmother, we always had a garden, we processed all our vegetables – I grew up helping in the kitchen and cooking," says Meyer. At age 15, he got his first job as a dishwasher in "a crappy chain restaurant", but he worked his way up. "I went off to college thinking I would do something else, but after two years I realised I loved cooking a lot more. I also played music in bands. I quit school, moved out here to the Pacific Northwest, became a vegan, started working in restaurants and touring with bands for three or four years until I realised I hated living in a van – it's romantic but no fun. That's when I bought into my first little restaurant, a tiny vegetarian cafe in southeast Portland. We were really busy and did very well. Then September 11th happened." He says he got really scared that society as we know it would break down and decided on some radical changes: he sold his cafe, got on his bike and started cycling through the US and Canada, earning a living by working on farms.

Two years later, he was back in the Pacific Northwest. Friends of his had a 10 acre organic small holding on Vashon Island, they raised a few heads of beef, chickens, turkeys, lambs, pigs and grew some row crops. With a newborn baby, they needed help on the farm. "While I was there I realised, 'oh, it's not meat that's bad'. I'm against factory farming, I'm not against animal agriculture. A few miles away there was a vegan farm that raised beautiful vegetables, but every week they had to bring in truckloads of compost to make it work. We didn't import anything because we had animals on the farm. But I knew that if I was going to eat animals, I would want to be responsible for them. I wanted to know where they were coming from and how they are handled. That's how, ultimately, I ended up here in butchery."

He left the farm and returned to Portland to work in a meat shop. In 2005, he and his business partner opened their first restaurant, Ned Ludd, followed by Grain & Gristle and Old Salt. Every week, they bought three sides of beef and three pigs, and butchered them in one of the restaurants with customers able to see how they worked. "The farmers we worked with kept warning us that with more and more slaughter facilities closing, they might no longer be able to supply us," says Meyer. The situation was so worrying that his business partner got in touch with an elderly couple who ran a small slaughter and butcher business, Revel Meat, near Canby. The facility was built in 1964, but the owners had no succession plan, they were struggling with ill health and found it hard to keep going. Together with one part time employee, they still managed to slaughter five head of beef and 35 sheep a week, but that wasn't enough to be profitable. Nevertheless, they still hung on to their USDA certification because they didn't want to let down the farmers who needed that service.

In 2016, Meyer started working at Revel Meat two days a week and in 2017, he and his business partner bought the facility. His goal was to work directly with organic and regenerative farmers and ranchers and provide market access for the meat they produced. "I knew all the chefs in Portland and I was sure that more of

Ben Meyer, the vegan turned slaughterman and butcher.

them would use local meat if they had easier access. Because at that point, if you were buying from a ranch, you pretty much had to buy a whole side of pork or a quarter of beef or a whole lamb, and for a lot of chefs that just doesn't work," says Meyer. Revel Meat would slaughter and process the meat and then supply wholesale to restaurants and high-end shops.

"This was at a time, when people in Portland and the States in general started talking about how beef is bad for the environment.

> Which is ridiculous because most of the land we raise beef on is not suitable for growing broccoli or grain. We wanted to showcase the fact that there is no single recipe for sustainability. Every piece of farmland needs somebody to pay attention to it. We wanted to show that some of the ranchers here have really unique operations and were doing right by their soil. Some of them had been doing it for five generations. Others had just started, but they all had pasture-based operations and cared deeply about the quality of life for the animals."

At present, Revel Meat slaughters and processes animals for 65 farmers and ranchers. They come from all over the Pacific Northwest, most are fairly local, but some drive for three or four hours. Meyer practices what David Zarling at NMPAN preaches: know how many work hours are available in a week and how many animals need to be slaughtered and processed at what price, in order to be profitable and pay at least a living wage.

Planning ahead is key. Some farmers bring eight pigs a year, others 40 head of cattle a month. Every July, Revel Meat sends out a calendar request, asking their customers how many animals they would like to bring for slaughter in the following year and in which months. They are now completely booked in advance for the whole of 2023, with detailed information accessible in a digital spread sheet. The bookings are reconfirmed by email a week in advance, if the rancher does not cancel immediately and turns out to be a no show, he will be penalised by having to pay the full slaughter costs. In discussions with each customer, the team establishes what cuts and produce they need. The information is entered onto a cut sheet. With the reconfirmation of the booking, farmers can also request changes – otherwise the existing cut sheet will be used.

Revel Meat sells between 20-25% of the meat wholesale, all other meat goes back to the farmers who have seven days to pick it up or they will be charged a $5 penalty per box per day.

Meyer takes us on a tour through the facility. From the office, we get straight into the cutting room – it's a Monday, the busiest day for cutting meat. A radio blasts music from a Latin station at full volume and everybody seems to be in a good mood. On the right side of the room workers are cutting lamb chops which have to go back into a cooler until the crew on the left side is ready to wrap and box the

Because of lack of space...

...meat often has to go back into the cooler.

meat. "It's an old building and we don't have the space to establish a proper line, efficiency really goes out of the window," says Meyer. Revel Meat slaughters beef, lamb, pork and goat, and it offers cut and wrap service and further processing into ground meat, along with sausages, marinated meats, smoked meats, bacon and ham. They can't do any ready-to-eat products, not even hot dogs, because cooked meat has to be kept separate from raw meat and they haven't got a second cooler.

When Meyer and his partner took over Revel Meat, the previous owner had not bought a new piece of equipment in over 30 years – "things were literally held together with duct tape". They took out a $400,000 loan and bought a new wrapping machine, new split and band saws, a new scalding tank for pigs and overhauled all coolers and refrigerators. Now, cooling space is the biggest issue. In a normal week, the team slaughters on three days: on Tuesday, 60-70 sheep (Revel Meat is one of very few facilities in the Pacific Northwest that still slaughters sheep); on Wednesday it's 20-40 pigs; followed by 24 heads of cattle on a Thursday. The demand for slaughter is extremely high and Revel Meat could process more animals if it weren't for the cooling space. Meyer shows us the cooler where the beef is aged for at least 14 days – we inch along the wall to get to the exit on the other side, the room is absolutely packed to capacity.

On a slaughter day, a team of six people works on the kill floor and one person brings up the animals from the enclosure in which they have spent the night. Farm equipment is used to constrain and stun or 'knock' the animals. Beef and sheep relax when they walk into the chute or 'squeeze', says Meyer, they've been in one multiple times on the farm, and they know that they are safe. Pigs are different, they are walked up in twos. The second one waits while the first one walks into the chute. As the chute is closed, a wedge at the back slightly raises the floor level so that the animal is pushed forward a little and can be safely stunned with electric tongs. For cattle, a pneumatic head catch is used, it is similar to a field head catch and familiar to the animals. Because of this device, the number of mis-stuns was reduced by 99%.

Humane slaughter is Meyer's number one priority.

Monday is not a slaughter day.

Cattle are used to being in a squeeze and a head yoke.

Every three months Meyer reviews with employees how animals are to be stunned.

Different bolt guns for animals of different size.

There are very few studies on the best practice for stunning an animal, says Meyer. Most research has been done by big meat processors such as Cargill or JBS, but these companies only slaughter Angus cattle with all animals being nearly identical in size and weight. "We handle all sizes and lots of different breeds." Meyer says they are in a similar situation to vets who have to be able to kill animals in the field with a hand-held captive bolt gun.

Meyer managed to get hold of a training manual for vets and together with the research published by Temple Grandin, the team worked out a detailed protocol for the humane handling of the animals – a process that took nearly a year. But now everybody knows exactly what to do when and how – from the time a trailer pulls up to the actual slaughter. "Every step, every detail is planned," says Meyer, "how to unload the animals, making sure they have enough water and the surfaces are clean so they don't slip and fall, how to lead them into the squeeze, how to knock them, what knocking device to use for which species, what weight ranges it can handle, what charges we use depending on their size, how we maintain the head catch…being a former vegan, it's my biggest fear that we do something wrong, that we mistreat an animal or that we don't notice if an employee doesn't care."

Humane handling starts with unloading the animals.

In and around the kill floor, there are posters with detailed explanations in Spanish and English as to where and how to stun different species, and the whole process is reviewed with the employees once every three months. Meyer has shared all their templates for their humane slaughter protocol with NMPAN, including information on what equipment they bought and how they financed it.

We step outside as a trailer pulls up. A farmer and his sister deliver pigs for slaughter. They have come from Mitchell in central Oregon, a good four-hour drive away. There are two groups of pigs on the trailer, both a mix of breeds, Gloucester Old Spot, Hampshire and Durac. Two Revel Meat staff open the trailer and attach the ramp for the pigs to get out. The first one steps out, head raised, sniffing the air inquisitively, the others follow. The men weigh the two groups and then settle them in their pens. They encourage the animals forward but never raise their voices and only occasionally snap their fingers, neither touches the pigs, the whole process is calm and measured, the animals are allowed to take their time and none seems in the least stressed.

The first pig makes its
way out of the trailer.

The others follow. No
need to shout or prod.

I ask Meyer about staff and training. He says they were able to hire
staff from a cut and wrap facility that closed down which meant
everyone had some experience. By now, there is a routine for new
staff, on which table they start and how they progress. Training
cutters on the job is hard but doable. The Butchers Union of America
used to have great training programmes, says Meyer, but with meat
being sold mostly through huge grocery chains, grocery cutters
"don't know where anything belongs on an animal, they just know
how to cut what comes out of a box". Making charcuterie needs
additional training. But there are still some jobs only two people at
Revel Meat can do – such as breaking down a whole beef carcass
according to the cutting instructions. When one of the two got
injured and couldn't work for six weeks, his partner had to step in
and do the work. The same goes for Meyer who will replace anyone
on the kill floor who is off sick. "That's time we don't have to work
on the business side of things and the scheduling goes out of the
window, it's really bad." In his opinion, the most difficult task is to
train someone to be a good slaughterman. Doing it safely, quickly,
efficiently and humanely is a really high skill, he says.

At present, Revel Meat has 26 employees and wages range from $18
to $30 an hour. There are no health insurance or retirement benefits.

"We constantly have to remind farmers that we have to raise prices
because we have to pay our people better. I tell them, if we don't

they will leave and I won't be able to handle their animals any longer. Everybody has to be in this together. Stop competing with supermarket meat. Explain to your customers what you do, that you raise your animals on pasture and why that matters. Yes, it is more expensive, but quality, taste and knowing that the animals were humanely handled is worth the money."

The pride of being artisan butchers.

Chapter 11

Skills needed: meat cutting, management and marketing

Alyssa Jumars works for the Washington State Department of Agriculture (WSDA) and her official title is Local Meat Marketing and Capacity Specialist. Her job was created as result of the pandemic because consumers suddenly wanted local food, meat in particular. Jumars focus is on identifying the challenges for small farmers and meat processors and figuring out how they might get better market access.

There are 140 state licensed but custom exempt small butcher shops in Washington. This means they are only allowed to slaughter and butcher animals for the person who owns the animal for their own consumption. Normally that would be the farm family only, but farmers and ranchers can sell a share of the living animal to consumers who will then be able to receive the meat, usually in the form of a meat box. A share can be as little as one-eighth of the animal.

Meat boxes have become extremely popular since the pandemic and there are certainly enough farmers and ranchers able to sell meat shares, but they struggle to have their animals slaughtered. Most slaughter facilities are fully booked and have waiting lists. And the slaughter capacity cannot be increased because of a lack of staff.

When we spoke, Jumars was still evaluating a survey conducted in June 2022. Preliminary results indicated that the biggest problem for small meat processors was finding staff: 95% of businesses said they had trouble finding trained staff and 75% found it difficult to hire even untrained workers; a further 75% said that three out of four people they were able to hire had little or no experience in meat processing.

To put this into context: Jumars defines a small butcher/meat processor as one with two to ten staff, slaughtering about 1,000 head of cattle or the equivalent in smaller animals per year. The existing 140 businesses serve about 20,000 small farmers, ranchers and livestock producers.

Beef is one of Washington's primary commodities, with calf rearing mostly in the north of the state and finishing operations in the south where the big meat packing plants are concentrated.

There are only about half a dozen USDA certified small plants, and ranchers will often have to do an eight hour round trip to deliver their animals for slaughter, and another such trip to collect the meat. That's why selling animal shares and custom exempt slaughter are of such importance. The Washington State legislature is therefore supporting the sector with a one-time grant of $3.5 million, says Jumars. Many butchers work out of old buildings with outdated machinery: the department received 110 applications and awarded 40 grants for measures to improve infrastructure, refrigeration, food safety and efficiency. Overall, the 110 applications amounted to $27 million, says Jumars, and to modernise and work efficiently, the sector would require further investment of around $70 million. And even if the money were available, it would not solve staffing and training issues which the niche meat processors have identified as their biggest problem.

To help with training, the WSDA is at present sponsoring access to the Range® Meat Academy programme. It is an online training programme which teaches whole carcass butchery. Jumars says workers either come from retail and know how to portion meat or they have worked in a big slaughter facility where they were trained to do one particular cut. Hardly anyone knows how to butcher a whole carcass. So far, 20 butcher shops have signed up 50 workers, who have one year to finish all the modules. If the feedback is positive, WSDA may use this course or elements of it to build a curriculum for butcher trainings in community colleges.

So, how does the Range® Meat Academy[123] work? Its creator and CEO is Kari Underly, a butcher and educator. Since the publication of her book, The Art of Beef Cutting[124], she is known among craft butchers not just in the US, but in the UK, too. The book is a beautifully illustrated step by step guide for cutting beef, working with ground beef, food safety, occupational health, strategies for profitability and a spice and flavour guide. The Range® Meat Academy builds on the book, covering not just butchering beef, but also pork, lamb and poultry. Underly has produced more than 10 hours of video content, and there are downloads, handouts and assessment sheets. At the end of the course, students get a certificate of completion. The course was rolled out in 2018 and the Range® Meat Academy is approved by the Illinois Board of Higher Education to operate as an online vocational training school.

Underly now lives on a farm in Indiana and I caught up with her on Zoom. I ask her why she decided to become a butcher. "I do come from a long line of butchers. My first vision of the world was through the lens of a butcher when I was a kid," she says. When she was little, she spent a lot of time in the family butcher shop and was allowed to help. "Dad put me up on a couple of milk crates and I'd stuff the ground beef down. I thought that was cool, I liked the tactile feeling, the smell, the freshness." When big box groceries started to expand into the suburbs, small butcheries were often no longer viable. Underly's father, too, lost the family business and worked in the meat department of a grocery store to provide for the family.

Nevertheless, Kari Underly was determined to become a butcher. She became the first female apprentice at Martin's Supermarket, a well-known grocery chain in the Midwest. As a female apprentice, there were "challenges", as she puts it, from being placed in the worst run stores to enduring crude jokes from "handsy" men. But Underly decided that by working hard and being "wicked smart", she could get through this. She qualified as journeywoman meat cutter through the company's apprenticeship programme, worked across all the stores of the chain, became a store manager and finally crossed over to the management side, helping with staff training, sourcing, planning, management and merchandising.

With limited career options at the supermarket chain, she decided to go back to what she really loved: butchery and training. The National Cattlemen's Beef Association commissioned her to develop new cuts that would add value to the meat. She created the 'flat iron steak', which "comes from the second most tender muscle in the whole animal". It's embedded in the chuck, which is mostly just used for braising, but "this one piece of excellent steak can be cut from it". She then spent time training butchers, chefs, restaurateurs and meat processors how to cut a flat iron.

For women, working in the meat industry is still extremely hard, says Underly who started to run training schemes exclusively for women, teaching them cutting skills that would qualify them for better paid jobs. For the trainees, the course could become quite emotional: one woman told Underly that she "gave her a voice, to have the strength and the power and the right vernacular to speak to these meat processors, because they would just shut her off: you are a woman, you don't know what you are talking about."

In 2002, Underly founded Range®. The idea was to connect farmers, abattoirs and chefs so that it would become easier to market meat directly. Adding value to meat takes well-trained butchers. Her dream always was to have a physical trade school in Chicago with "students looking sharp, giving them confidence, being proud of who they are and what they do, changing the image of the industry". She had a lot of support for the idea, but the bureaucratic and financial hurdles were just too high.

But some of her goals could be achieved with online training. A mentor at the DePaul University, Department of Education helped her set up a curriculum and encouraged her to getting licensed as a vocational trade school, a process that took two years. The course is strong on identification, says Underly, learning where the muscle groups are and how they connect to the joint, understanding where the separation points are and where the cuts have to be made. Meat cutting is, of course, a physical skill. The course works well in a work environment where students can practice what they learnt, says Underly, which is exactly how the Range® Meat Academy course is being used in Washington.

Kari Underly is a craft butcher and expert meat cutter. What she cannot teach, not even in theory, is how to slaughter an animal. There are hardly any opportunities to learn it, she says, some Land Grant Universities do teach butchery and slaughter, but university students are unlikely to end up working as butchers and slaughtermen.

In the Pacific Northwest, small meat processors in the 'niche meat sector' have long been worried about staff and staff training. The startling results from the WSDA survey only confirm how big the problem is. The Northwest Meat Processors Association, NWMPA, has decided to change this and launch a full three-year apprenticeship programme. Troy Wilcox is the NWMPA's executive director and programme manager. We meet him in Chehalis, a town exactly half-way between Seattle and Portland.

Wilcox is not a butcher but works in the packing industry, a job that brought him in close contact with butchers and cut and wrap facilities across the Pacific Northwest. Eight years ago, he joined NWMPA and was nominated for the board two years later. Everyone agreed that training butchers better was a priority, he says. At the time, only the neighbouring state of Montana had an apprenticeship programme under the auspices of the Montana State University. It could serve as a blueprint for a similar programme in Washington and Oregon. In both states, the Department of Labour and Industries approves, supports and administers apprenticeship programmes and the new butcher apprenticeship scheme would have been given the go-ahead in 2019, had it not been for a last-minute intervention by two Unions.

Rather than fight the Unions in court, NWMPA members decided to launch the scheme on their own with the executive board providing the necessary oversight.

The NWMPA's three-year programme is open for apprentices from Washington, Oregon and Idaho. Butchers and meat processors can apply to become 'training agents'. They must be able to teach everything, from slaughter and butchery right through to retail. The NWMPA apprenticeship handbook specifies the details: training on safety, 200 hours; pre-inspection of livestock for slaughter, 500 hours; slaughter of livestock, 600 hours; sanitation, 600 hours; preparation of merchandise, 400 hours; beef, 1300 hours; pork, 1,400 hours; smoking and curing, 600 hours; poultry, 200 hours; offal, 200 hours.

The training agent has to record the hours and report to NWMPA once a month. The practical training is complemented by textbook learning – successfully completing the Montana State University butchery online modules is a requirement. And to become a fully qualified butcher journeyman or journeywoman, the apprentice, at the end of three years, has to pass a final exam.

From the start, the apprentice has to be paid $16.75 per hour, 67% of an average journeyman's wage. The minimum wage in Washington is $14.25, a journeyman earns between $22 and $25 (wages in rural eastern Washington are lower). There are USDA grants available that pay training agents up to $7,000 a year to make up for the higher wage they need to pay.

The apprentice has to pay for the Montana State University course[125], which costs $945 when bought as a package, or $1,560 when modules are purchased separately. The money is refunded by the training agent once the apprentice has completed the course successfully. The training agent is reimbursed by NWMPA once the apprentice passes the journeyman exam. "Everyone has financial skin in the game," says Troy Wilcox.

So far, four training agents could be recruited and at present there are two apprentices, a third one has already qualified as a journeyman. Wilcox would like to add more training modules such as accounting, tax law, understanding benefit packages and team leadership. But those plans may have to wait. "Right now, we need to recruit more training agents and spread the word about the apprenticeship program, we could easily accommodate 10 to 20 trainees at any given time".

Ben Meyer from Revel Meat is one of the NWMPA board members and thinks the apprenticeship training programme is fantastic. In his opinion, the biggest hurdle might be the higher entry level pay for the apprentices. It is subsidised and the apprentices should earn decent wages from the start, but other employees may consider that unfair and decide to leave. He says many businesses will not be in the position to raise the overall wage level because they are already squeezed and do not want to increase costs to farmers.

David Zarling from NMPAN believes that decent wages are crucial for the whole industry. He says that across the US, apprenticeship programmes are highly competitive. "My little brother is 17 years younger than me, and he is trying to become a linesman to work on power lines; he's a straight A student but he can't get an apprenticeship because there are so many people applying. But the meat apprenticeships are staying vacant because people know they'll work their butts off and make 20 bucks an hour and you can't live off that anymore." Zarling believes that butcher training is not the issue: "Right now, the government is talking about workforce development. They say that we need more training programs for butchers. And I will die on this hill: there are enough butchers out there. They don't want to butcher because there is no money in it. The workforce we need to develop is business owners, so that they can understand how to run a viable business so that they can pay people living wages. Once those jobs exist, we'll have people banging on the doors of meat cutting programs."

And for Zarling, the problem goes even further: food is too cheap and that means everybody in the niche meat sector is being squeezed – the owners of small meat processing businesses such as Cured Processing, The Meating Place or Revel Meat, butchers and slaughtermen as well as farmers.

"In a perfect world, every county would have multiple mobile slaughter units, one or two raw processors and one value added facility like us." Zarling says the argument could be that every stage adds costs, but it takes time and effort to make a quality product and that needs to be paid for. "No one, in this country at least, can put a 16 oz rib-eye into a modern package for less than a dollar a pound. That is the cost if you pay people a living wage. Some facilities on the East Coast still charge $0.65 a pound to process beef. They are either losing money every day they operate, or they don't pay a living wage. That's why we have this tremendous shortage in the meat processing industry in the US, nobody wants to work in an industry that doesn't appreciate them."

Farmers end up in a similar trap of self-exploitation or trying to squeeze others in the chain to survive. When their production costs are increasing, they will demand very low prices from processors, which translates into a lower and lower quality commodity product, says Zarling. His advice: "Get organised!"

"So many farmers try to do everything, and they get crushed. It shouldn't kill you to be a farmer. By pooling your resources you create enough margin that you can share your marketing workload. If you had an aggregated brand, let's say all farmers in this county got together and they aggregated all of their 50 head herds to a 2,000 head herd, they can schedule multiple animals at a time, cutting the same way, marketing directly to consumers – they could help each other and keep their cost of production low. Maybe

> then one producer in this farmer coop or organisation is able to take on that duty saying: I'm going to be the lead marketer but I'm not going to have anything to do with scheduling the animals for slaughter because this person in our cooperative will do that, I'm not going to deal with logistics because this other person handles that. That's the kind of cooperation that's needed. Pooling the resources is where the time and the energy and the margin come from, that allow you to actually have a marketing programme."

David Zarling is not the only one who believes that better cooperation is the way forward. I had recently read Bet the Farm, Beth Hoffman's account of how she and her husband John started farming in Iowa when she was in her 40s. To say that moving from California to the Midwest and being a novice farmer wasn't exactly easy, is an understatement. In one of the chapters, she describes meeting fellow farmer Nick Wallace[126]: "We were talking on the phone about the idea of Iowa ranchers working together to market and sell beef. I mentioned creating a cooperative, when Wallace posed an interesting question. "But what do farmers think when they hear the word 'co-op'?" Wallace asked me. "Do farmers think it's a good thing?"

"John and I met Wallace at a Practical Farmers of Iowa[127] gathering of meat producers, a group that came together to see if we could figure out better ways to sell our products. Many farmers in the room marketed directly to consumers via websites and farmers markets – mechanisms that sometimes generated higher per-pound sales but often left farmers exhausted. Others sold to big processors via the sale barn or through feedlot contracts – a much less tiring way to offload cattle but one that produced profit margins so small they barely covered the cost of raising the animals, if they did at all. It was there that Wallace presented his ideas for a meat

Promoting grass-fed beef is what Nick Wallace does.

company. The business would purchase animals raised without chemicals for above-market rates, he said. Then he would sell the products via the internet to customers in nearby cities, like Omaha or Chicago", writes Hoffman.

I was intrigued and contacted Wallace, who immediately agreed to us visiting him on his farm in Keystone, a tiny community about 150 kilometres northeast of Iowa's capital, Des Moines. The farmhouse, barns and storage units are newly built and modern, but with the grazing red and black Angus dotted across a recently harvested maize field, it's a picture book idyll. Nick Wallace has been selling meat from the family farm for nearly two decades, but "what we are doing now is so much bigger than just one farm," he says, as we sit around the conference table in the top floor office above the storage and packing area. In spring of 2022, he renamed the business '99 Counties', which in the Midwest is often used as a synonym for the State of Iowa. The website[128] went up in October, and on it, Wallace clearly defines his vision and mission:

> "99 Counties is a whole animal brand connecting consumers to local family farms. Through fair pricing, supply chain coordination, marketing and delivery, we enable farmers to reclaim their independence, focusing on farming in a way that's better for land, animals, people and the planet. (...) We believe that regenerative agriculture holds the key to solving many of the problems we face today. By supporting farmers to transition to regenerative agriculture, we're combatting climate change, while rebuilding farm town communities and providing families with high-quality food."

Farmers are not paid enough, in particular not for grass-fed animals, says Wallace, and "they can't do it all, farming, marketing, logistics, delivery...how do you have a family, too?" That's why so many have turned to feed lots, where animals will reach slaughter weight in record time, he says. The meat is sold cheaply in big box stores, easily accessible from highways and interstates. "This system has made it possible to bypass small rural towns, which as a consequence have died. Customers are now used to cheap, low value produce," says Wallace who sees 99 Counties as the beginning of a movement to change the regional food system not just in Iowa, but in the Midwest. For the physical and mental health of farmers and ranchers, to once again provide nutritious and nutrient dense food, and for the benefit to the environment, "we need to go back to the diverse farming systems we used to have, but with the technological means available today we can do marketing and distribution via the internet and through social media".

Trying to change the whole food system, one meat sale at a time is an ambitious goal, but as we talk about his early career and growing up in rural Iowa, it becomes obvious why Wallace aims high. At the age of 20, he was diagnosed with Hodgkin's lymphoma, a cancer that has been linked to pesticide use. In 2018, a jury in San Francisco awarded Dewayne Johnson, a groundskeeper who had to use glyphosate on a regular basis, a record sum of $289.2 million in damages. In his late teens, Wallace was a keen baseball player: he believes that the general contamination with pesticides in rural areas and the exposure to glyphosate used to maintain baseball fields are the likely causes for him developing Hodgkin's lymphoma.

Even after 25 years, regaining his health and remaining cancer-free seems almost miraculous. When he was diagnosed, his parents decided to try and eliminate pesticides from their diet as much as possible. The whole family switched to eating organic food and the farm has been certified organic for many years. Nick Wallace initially left Iowa and farming, for a career in corporate insurance in Chicago. After a few years, he knew that finance wasn't right for him. He moved to Colorado, went to culinary school and started working at a high-end restaurant. He was always curious where the food came from, and the answer 'from a distributor' wasn't good enough for him, he says. He still remembers the day he decided to return to Iowa and the family farm: "It was on Christmas morning, 8 am, I was stressed off my feet making omelettes for rich skiers in a restaurant, and it was the first time I did not spend the holidays with my family. I talked to my boss about it, and she said, 'get used to working nights, weekends and holidays, or go into canteens and corporate food'. I wanted neither."

Back on the farm, his father and sister had just launched a farm website, but it wasn't very successful. For want of anything better to do, Nick Wallace took it on and started to sell the farm's beef online by sending out emails to everyone he knew with information on what was on offer. Word spread and Wallace started buying in meat from other farms, adding pork, turkey and poultry. Not all meat was certified organic, but animals had to be raised on non-GMO feed, and not in confinement. Initially, there were drop-off points where he

On a beautiful fall day the cows are gleaning in what was a corn field.

A proud Angus.

would be stationed for two hours at a time for customers to pick up their orders. Until 2017, Wallace was able to continuously increase his customer base but then everything just went flat, he says. "Groceries had realised that consumers liked grass-fed beef and were willing to pay more for it. By 2017, they had started bringing it in frozen from New Zealand and South America, selling it much cheaper than we could," says Wallace. Rather than using drop-off points, he decided to deliver meat in a 100-mile-radius, which includes the cities of Des Moines, Cedar Rapids, Cedar Falls and Ames. He also rented a small warehouse in a Chicago suburb, where a driver makes deliveries three days a week. 99 Counties builds on this foundation. At present, Wallace cooperates with 23 farmers – Beth Hoffman being one of them. He buys animals at the farm gate and for a premium – "farmers have to make money too". 99 Counties arranges for transport to one of five processors. Wallace has invested in warehouse and freezer space, which means processors can return all produce to the farm, where orders are then assembled for delivery.

Customers order online. For an annual fee of $99, customers can become members which means all orders over $75 are delivered free of charge; for everyone else the delivery charge is $15. Animal carcasses are used 'nose to tail', and on offer are 60 different meat products. So far, 99 Counties has a customer base of about 1,000 and on average it gets 700 orders a month. When we spoke, Wallace had just been in talks with a cooking club in Chicago, and chances are that they may come on board with regular orders. At present, he is likely to market 300 to 400 head of beef a year, the monthly average stands at 30. He hopes it will be 30 a week by 2025, and 1,000 to 1,200 hogs by the end of 2024.

Farm size may become a problem – there aren't enough small and medium size farms anymore, says Wallace, you either have really big operations or 'boutique farms' raising five animals a year. He tries to meet farmers where they are, they have to supply a minimum of five or six head of cattle, but he prefers to collect 10 head of beef at a time (or 25 to 30 pigs or 1,000 chickens). Wallace aggregates pick-ups – a full trailer will hold 30 animals, the minimum load is 20 animals, everything else is not economically viable.

Wallace defines as 'regional' any place that can be reached within five hours – with another five hours for the return journey, that makes for a long work day, but it would allow for meat sales to customers across the Midwest including cities such as Minneapolis, Omaha, St. Louis, Kansas City and Chicago. The biggest challenge is to get consumers to realise that their food dollars count, and that it is worth it to spend a bit more because "meat from grass-fed animals not only tastes better, it is good for your health," says Wallace.

He admits that most of his customers are well-heeled. For 99 Counties to be viable and the farmers receiving a fair price for their animals, he has to sell beef for about $12 a pound, pork for $10 and chicken $8 – a US pound is 453g, the price per kilogram therefore would be roughly double that.

At present, the priority is building the customer base, increasing meat sales and thus providing market access for more small and medium sized farms in Iowa. But slaughter is a concern. Wallace is working with small, independent slaughter and processing facilities, but the animals often have to be transported over fairly long distances. He, too, believes, that it would be ideal to slaughter animals on farm or as close to it as possible. "In a perfect world, there would be five to fifteen young people who have a mobile slaughter unit and a team to run it. We would commission them and

tell them: these are the farms with X number of animals to slaughter. The butcher teams would work out their own schedule and deliver the chilled halves to us for value adding."

For David Zarling from NMPAN, this is a model that can work but it's not the only option. He believes that there is no 'one-size-fits-all' solution, it's all about cooperation, aggregation and costing. "Everything should be costed beforehand," says Zarling, not just in meat processing, but in farming too. "Farmers need to decide how much revenue they need to make from meat sales, and that is the basis for what a co-op model needs to look like." It can make sense for a co-op to own an MSU, organise the slaughter schedule and deliver the carcasses to a cut and wrap facility with the option to get value added products such as charcuterie. It can also make sense for a cut and wrap facility to slaughter and cooperate with a value adding facility – the model of the Island Grower Farmer Co-op.

On-farm slaughter is only financially viable if the farmer has a sufficient number of animals to slaughter or acts as a host for the MSU and an aggregation point, to which other ranchers in the neighbourhood can bring their animals on the day or the night before. When he worked on an MSU, there was a flat fee of $400 just to set it up and operate it for the day.

Zarling distinguishes clearly between being profitable and being efficient, for him it's about balance: work culture – from the humane handling of animals to earning at least a living wage – matters just as much as efficiency. "Of course, we have meat processors in Washington where 2,000 head of cattle are slaughtered and processed in a day. The process is broken up into individual steps and everyone gets very efficient at what they do. A gutter may well be able to gut and split an animal in 20 seconds. And that's where we lose the human aspect. That's where it gets really mentally taxing and difficult. But if you have a facility that has a kill floor with five stations and people rotate because everyone is just as proficient as someone else, that's an awesome company culture." It's a model that can work in a small abattoir such as Legau in Germany or with an MSU. Travis Stockstill, the manager at the Island Grown Farmers Cooperative rotates teams on the MSU, the butchers who are not in the field in any given week will be processing meat at the cut and wrap facility in Burlington.

With cooperation, aggregation and costing, a value added facility such as Cured Processing can deliver to every farmer in a group what he would be unable to deliver to the same farmers if they were to come to him individually. "There used to be 20 different producers who all wanted me to do 200 pounds of snack sticks at a time, or sausage or whatever – I told them, I can't do this. It's too hard to organise and schedule all your separate stuff, keep all your batches separate, my smoker runs 800 pounds at a time and it costs me $125 every time I turn it on. That's the utility cost of this machine. So, if I spread that cost over 200 pounds, I'm going to lose money, I'm going to pay you to take your product back. If I have 20 producers and they aggregate the meat and bring me what I need to do a minimum run every single time, I can make sure that they get their product back with their label on it and a margin that keeps our lights on."

With good planning, processing and value adding, facilities can also accommodate organic farmers and process non-GMO or 'grass-fed only' meat. There is a segregation protocol, says Zarling, a sanitation cycle will be run at the end of every workday in any case, if organic meat is processed first thing in the morning, the only other requirement is the need for separate storage. It's all about scheduling. And with cooperation throughout the chain, from farmers to slaughter to processing and value adding, it is possible to slaughter animals humanely, on the farm or close to it and for everyone to have a share in the profits, too.

Part 4

Humane, local and profitable meat in the UK

Chapter 12

Lessons from the EU and the US and how to implement them here

The examples from the US and an EU country such as Germany show that given the right setup, it is possible to slaughter animals on farm or in close proximity, minimising transport time and stress. It's possible to slaughter animals humanely and for farmers and butchers to stay profitable nevertheless. It's not been easy to get to that point, neither in the US nor in the EU: it took the combined efforts of farmers, ranchers, vets, butchers, chefs and consumers, and, in the EU, it took regulators to finally implement the necessary changes.

As Shinn and Pledger write: "Family farms and ranchers and related livestock enterprises can flourish, especially if adequate, appropriately sized slaughter and processing facilities are developed. Food supplies closer to home – to everyone's home – would require resources and opportunities for small- and medium -sized businesses to succeed. It would also mean prioritizing decent livelihoods and working conditions for people employed in local and regional food systems"[129].

So, what are the options? Mobile Slaughter Units, MSUs, have been seen as holding a lot of potential, both on the Continent and in the US.

In Sweden, MSUs have been used for several decades now. Hälsingestintan[130], a Swedish company, has developed a system with two trailers, one for slaughter, fitted with an independent electricity and water supply and a second one for cold storage. Kometos[131], a Finnish company, works with a slightly different design: independent modules contain slaughter facility and cold storage. The modules can be placed on a semi-trailer and transported from one farm to the next. Initially, the units were developed for the slaughter of reindeer in Finland, but they are also suitable for cattle. MSUs from both companies are now in use in France[132], too.

In 2020, the Scottish government published a report[133] on the viability and sustainability of mobile abattoirs. The study looked at two different models: an MSU driven to an individual farm on the day animals were to be slaughtered and a slaughter hub/docking station at a farm, or a mart, that could be used by farms in the vicinity with farmers transporting their animals over a short distance. "The review identified that the most practical and preferred solution was the docking station approach and a cost benefit analysis (CBA) was then undertaken based on MSUs using a docking station approach, which could for example, involve a unit driving to the following types of location, which already have much of the required infrastructure in place (lairage, drainage, power, water): an auction mart; an industrial unit; and/or a farm".

To assess the demand, the researchers conducted a survey. Of the 600 respondents, 90% said they would support and use an MSU. The problem though is the price tag to an MSU solution: "The cost for authorising and maintaining a service, in terms of compliance costs associated with approving an MSU, waste management and veterinary and meat hygiene inspections, has been shown to be a very small part of the overall costs of any future MSU. The most significant costs are those for staffing, waste disposal, maintenance (of the capital equipment) and debt financing. The operational models considered in the cost benefit analysis require docking station locations to form part of a future MSU service, with auction/livestock marts, farms, and industrial units potentially viable places, with chill units installed for hanging carcasses. The operating models considered for a future MSU service included these as stand-alone businesses providing butchers, meat processors and farmers with carcasses (e.g. sides, quarters). In terms of the types of MSUs that would be required a number of options were considered and a cost was used that allows the kill, evisceration, cutting (quarters and sides) with limited, temporary chill facilities in the trailers themselves. This requires waste to be left at the docking station locations, in secure containers, with collection by a registered carrier then taking place without undue delay (likely to be in line with fallen stock timelines). The capital cost associated with this model is between £800K and £900K".

The costs of running an MSU are high, but as the examples from the US show: under certain conditions it still might be the perfect solution. There are several dozen islands of the western coast of the northern state of Washington which are connected to the mainland by ferry. The Island Grown Farmers Cooperative regularly stations their MSU on a designated farm on one of the islands and farmers bring their animals for slaughter to that location. The carcasses are then taken to and processed in a facility in Burlington which I visited. The situation is very similar to the islands off the coast of Scotland. Few islands still have a slaughter facility and even a small abattoir may not be financially viable. But transporting animals to the mainland by ferry is not just stressful, in bad weather the journey may be long and hazardous. The deployment of an MSU would need to be well coordinated, but with sufficient planning it could be stationed on an island whenever there is enough demand. The experience in the US also shows that the operators of an MSU should work with a network of designated, fully equipped host farms. Animals from neighbouring farms will still have to go on a short journey, but a host farm will help to save time and costs. Long term, running an MSU can't be a loss-making enterprise.

Large MSUs such as the ones run by Mike Callicrate, the North Cascades Meat Producers Cooperative and the Island Grown Farmers Cooperative don't make sense on the British mainland: they are far too big to navigate narrow country lanes and narrow farm gates and the operating costs are high. The animals will be slaughtered on farm, but if there is a small abattoir nearby, there is no need for an expensive MSU.

The situation is different for sheep. Jeff Rogers used readily available parts and components to build a small MSU for slaughtering sheep. The cold storage unit is detachable. Both, slaughter and cooling unit, are mounted on a trailer that can be pulled by an SUV or pick-up. Rogers estimates that it might cost around £140,000 to build the unit today, but because it can be transported very easily from farm to farm, a group of farmers could share ownership and use. It could be a great solution for sheep farmers across the UK, but for that to happen, the necessary regulatory framework would need to be put in place. This could be done immediately and without the necessity for any kind of study or evaluation simply by adopting the relevant paragraphs of the amended EU regulation which is based on lengthy consultations and evaluations by veterinarians, food safety and food hygiene experts. And the experience from Germany shows that since the introduction of the amendment in 2021 no problems or issues have arisen.

The adoption of the amended EU regulation in the UK would also pave the way for the most humane form of slaughter of all: on the farm slaughter and killing animals within the herd. Mechthild Knösel, the dairy farmer near Lake Constance exemplifies what is possible – by learning to use a stun gun she is now able to prepare an animal in such a way that it will not even be spooked by an unfamiliar movement and the appearance of the butcher, a person it does not know. Such preparation would be unnecessary, were Knösel able to shoot the animal in the field within the herd. At present the regulations specify that a rifle shot can only be used for animals that are on grass year-round. There is no conceivable reason why animals that are housed for just a short period of time in winter cannot be killed within the herd too.

Apart from the expense for training and the steep fees for the presence of the vet and the butcher, the costs for equipment are low. All that is needed is a relatively simple container, a 'slaughter box', in which the carcass can be transported to an abattoir or butcher to be de-hided, gutted and processed. The box has to have a grid on which the animal can be placed and a container that fits underneath to catch the blood. The MSB® II[134] was developed by a not for profit association and costs about Euro 10,000 which amounts to roughly £8,500. A group of farmers could easily share ownership and use.

Again, if the EU framework were to be adopted here, this type of humane slaughter could be made possible in the UK immediately. The protocol for slaughter has been tried and tested and the equipment is readily available.

During my research in Germany and the US it became clear that "small abattoirs" are certainly not just defined by the number of animals slaughtered per year. Location and land ownership matter.

Business models vary: while in some instances one or more farmers took the initiative to build and run an abattoir, there were as many examples where butchers expanded their business and branched out into custom slaughter. Which course of action farmers and butchers took depended on a number of circumstances, including location, time and financial means available. Therefore, there is no general lesson to be drawn, but each of the cases described may be just the right fit or contain the key elements of a solution for a particular group of farmers or butchers in the UK.

By investing into a small abattoir on her farm, Judith Wolfarth created an ideal scenario: not only can the pigs be slaughtered humanely, they don't have to leave the farm, they are with other pigs, on grassland with trees that provide shade and experience no stress until the very last seconds of their lives; most will never 'know' what happened. Wolfarth has also invested in processing equipment for making sausages as well as high end charcuterie such as salamis and air-dried, aged hams. The unit could run profitably if neighbouring pig farmers could bring their animals for slaughter, but that's not possible for legal reasons: in Germany, an abattoir that has been built by a farmer on his or her land can only process the animals of that particular farm. An exemption to that one rule would make all the difference – for better animal welfare on neighbouring farms and for Wolfarth's finances. The facility itself would not have to be altered, it simply would be used efficiently.

While Judith Wolfarth direct markets her products to private customers and restaurants, the farm set up by the construction company Lindner Group serves a much wider market: it supplies the company's canteens, several restaurants and hotels, there is an on-farm restaurant and shop and a mail order business. As a result, the abattoir and butchery need three full time staff and there is an apprentice now, too. As a company that has nothing to do with food production, Lindner Group may be unique in nevertheless prioritising good food for their employees and investing into the local (food) economy. But it shows what is possible with market access. There is no reason why companies in Britain can't invest into a local area in a similar fashion.

None of what the Lindner Group farm does would be possible without well trained staff and here, too, the company invests heavily: there are apprenticeship programmes not just for butchers, but a number of other professions too, including in agriculture, sales and marketing. One farm enterprise can have a huge positive impact on a rural community in a rather remote area. The abattoir is a small but vital piece in this setup.

The abattoir in the small town of Legau is owned by a farmer cooperative and run by butchers, two of whom are on the coop's board of directors. The facility is financially viable, because the butchers buy animals from members of the cooperative for slaughter, they process the meat, produce dozens of different sausages and other products, all of which is then sold through a busy butcher's shop and supplied to several other outlets. The schedule leaves time to custom slaughter and butcher animals for members of the cooperative who sell the meat online or directly through a farm shop, keeping the value addition as close to the farm as possible. The Legau abattoir is particularly interesting because most of the coop members are dairy farmers. The animals, mostly local breeds, well adapted to the climate, are out on grass for most of the year and housed only in winter. High yielding dairy cows such as Holsteins are usually kept in confinement, the enormous amount of milk they produce takes such a toll on their body that they often have to be slaughtered when they have had only one or two calves. Usually, their meat can only be used for hamburgers or goes to the pet food industry. Dairy farmers like those in the Legau cooperative, who work with a grass-based system and more robust breeds, will normally not see milk yields drop until the animals are aged seven or older, and the meat of these dairy cows has an excellent quality and is sought after by chefs. Because they have access to an abattoir the farmers have an additional high value product for sale.

Only recently, Dan Barber, the chef and owner of the two Michelin star restaurant Blue Hill at Stone Barns said that when done right, dairy cow meat was superior to Wagyu. "There's a wealth of untapped potential in this meat"[135], Barber told Ambrook Research. "Retired dairy cow meat takes what we already know about the benefits of grass-fed beef one logical step further. (...) It's possibly the most regenerative and resource-conscious beef out there". The article concludes that "selling cull dairy cow meat to upscale restaurants does unlock new revenue generators for dairy farmers". There are high end UK butchers who sell dairy cow steaks at top prices, but for dairy farmers to tap into that market they need to have access to an abattoir. The model of the farmer cooperative owned abattoir in Legau is an example how this can work.

And there is another lesson to be drawn from the small abattoir in Legau: the butchers slaughter on only two mornings a week, typically five or six pigs and one or two cattle. Out of their working week they spend mere minutes actually killing an animal. The overwhelming majority of their the time is spent on processing, preparing cuts, producing products such as sausages, dealing with paperwork, managerial tasks, visiting farms to buy animals. The act of stunning the animal and bleeding it out is a necessary part of their work, which they do with as much care for the welfare of the animals as possible. Humane slaughter is their absolute priority and their work environment, the Tuesday and Friday mornings in this small abattoir, allow them to never lose sight of their own humanity and what it means to take the life of an animal for food.

In the US we visited just one farmer owned small abattoir, but that, too, held an important lesson: the butchers at the facility in Bowden, North Dakota, can slaughter not just pigs and cattle, but also bison. For that a special holding pen with extra high walls had to be installed and the chute modified. In future, more bison will likely be delivered for slaughter: Environmentalists have long realised the importance of bison for prairie grassland, the excellent qualities of bison meat are becoming more widely known which incentivises more ranchers to raise them, and a pilot scheme by the US government promotes bison meat in tribal communities in the Dakotas. The lesson for small abattoirs in the UK? Rare breeds are well adapted to different environments and particularly well suited for the creation and management of habitats. Many have strong and often long horns and their thick, woolly coats protect them from wind and rain. Most abattoirs cannot or will not deal with 'non-standard' animals that are particularly big or small, have very thick coats or unusual horns. Although the value of such breeds as 'land managers' is increasingly being recognised, at the end of their lives, the length of their journey to the abattoir is constantly increasing. Maintaining existing and creating new, small facilities that are equipped to slaughter rare breeds is of vital importance.

A number of examples from Germany and the US show that the need for farmers to get their animals to slaughter in order to direct market the meat provides new business opportunities for butchers, too.

When Gero Konrad took on the family butcher shop in Lenningen near Stuttgart he had to make a decision: start buying in meat for his produce or build a new abattoir to continue slaughtering. He decided to invest in a new facility next to the shop, a financial commitment he could only make if he expanded into custom slaughter. Stuttgart is an industry hub, many farmers have found well-paying jobs and now farm only part-time, keeping just a few animals. Other farms have diversified and are certified organic. To cater for both groups, the new abattoir is certified organic and licensed to slaughter cattle, pigs, goats, sheep and lambs. 60% of the business will be custom slaughter for farmers in a 50 kilometre radius, 40% will be animals slaughtered for the meat supply of the shop. In the UK, butchers like John Mettrick have employed the same model and many farmers were devastated when he was forced to close down the abattoir. The example from Germany simply proves that there is real demand for butchers who offer custom slaughter services. In the UK, regulatory change would be a crucial first step. Veterinary oversight during slaughter makes absolute sense. But the need for the constant presence of a vet at any small facility slaughtering a few more than 1,000 animals p/a is arbitrary and makes no sense – not for food safety and hygiene, nor for more the humane handling and slaughter of animals.

Klaus Bonkhoff a third-generation butcher in North Rhine-Westphalia has invested in a small MSU that can hold up to three beef carcasses. His on-farm custom slaughter is one of the mainstays of the business despite the fact that the costs of veterinary oversight are prohibitively high. Bonkhoff argues that many more farmers would choose this most humane way of slaughter if fees could be reduced, subsidised by the regional government or waved altogether.

Butchers who offer custom slaughter services in their abattoir or by operating an MSU are vital for building or expanding a local food chain. Not only do they link farmers and customers, very often they provide market access and create skilled jobs in rural communities that otherwise lack them.

A good example it The Meating Place west of Portland, Oregon. The business custom cuts meat, produces sausages and products such as hamburger patties, the shop offers a huge selection of cold cuts, ready to cook meats, roasts, steaks and ribs and the cafe has a busy breakfast and lunch service. Recently, The Meating Place invested into a medium size MSU, the team can now offer a custom slaughter service for farmers and the meat sold in the shop no longer has to be bought in. In future animals will be bought from farmers directly, slaughtered on the premises and processed at The Meating Place – which means the products on sale in the shop and served in the restaurant will be fully traceable. At present, the business employs 60 people, all able to work in a number of different roles. With the expansion into on-farm slaughter, more staff will likely be needed.

Revel Meats, too, is a company that links farmers to customers. It was Ben Meyer, the chef and owner of two high end farm to fork restaurants with a policy to be as sustainable as possible, who realised that local farmers and rancher would no longer be able to supply him with meat if they couldn't find an abattoir to get them slaughtered. And it was his commitment to creating a local and sustainable food system that had him save a small, family run abattoir from closure. The unit now serves 65 farmers who take 75% - 80% of the meat back to sell directly, the rest Meyer sells wholesale.

Lessons can be drawn from each of the case studies in this book. There are various models that make sure animals are slaughtered humanely and stress free. In the US and in the EU/Germany procedures and legal frame works are in place that could be adopted in the UK for these options to be available in the UK, too. MSUs, the use of slaughter boxes and small abattoirs increase the chances that farmers and ranchers can direct market their meat. Small, local facilities bring jobs to rural areas. Each and every solutions farmers, butchers and chefs found contribute vital elements to the problem we are facing: a sustainable, local food system that provides healthy food to customers, environmental benefits and a livelihood to farmers, butchers and the people they employ.

The most important take away therefore is that we need to get to a linked-up, humane, efficient and profitable system.

Well-trained butcher-slaughtermen are central. Small abattoirs are part of the solution, but it's not about brick and mortar structures – it's about the people who run them. Britain is so short of trained butchers and slaughtermen that recruiters are now bringing in workers from as far away as the Philippines to work in large slaughter facilities. Small abattoirs need craft butchers. In small abattoirs, MSUs and independent butcher shops, small teams of well-trained butchers need to be able to perform all necessary tasks, from slaughter to cutting, processing, creating value added products such as charcuterie, to sales and marketing.

Three-year apprenticeship programmes will teach all these skills which gives craft butchers the option to specialise while still being trained to work across different stations. Whether training is as formalised as in Germany, or trade bodies come up with a curriculum as in the Pacific Northwest – the skill set students acquire has to be broad. The image of what it means to be a butcher is slowly changing. The average age of farmers in the UK is 59, but more young people want to come into the industry because they want to grow food and are interested in health and nutrition, they care for the environment, they don't want to work in the city but be connected to nature…Once young people see the choice of career options for well-trained craft-butchers the profession will likely become a lot more attractive. In the UK, there is at present no clear path into the industry, 'craft butcher' is not an official title based on clearly defined standards that have to be met before it is granted. To achieve sustainable meat processing in the UK, the issue of butcher training has to be addressed.

The same goes for working conditions and wages in the industry. "It shouldn't kill you to be a farmer," David Zarling from NMPAN said. Nor should you risk your health butchering for minimum wage. The meat food chain – from slaughter to ready to sell meat products in a farm shop, online or in a meat box – needs to be profitable for everyone involved. From farm to slaughter to processing – there has to be a business plan, you have to know what your bottom line needs to be, says Zarling. In his opinion, the key to profitability is aggregation and cooperation. Organisational structure and ownership are secondary: whether an MSU is owned by a farmer co-op and butchers work as contractors, or whether a meat processor owns an MSU which farmers can book – the appropriate model is the one that works under the given circumstances.

NMPAN, the Niche Market Processor Assistance Network, is a University of Oregon Extension-based community but the work is not limited to the US: butchers, farmers and processors can access the huge back catalogue of webinars, videos and fact sheets. Together with the American Association of Meat Processors (AAMP), NMPAN has an eight-week course for small niche meat processors to get an introduction into best business practices. The curriculum ranges from business planning to financial management, human resources and marketing and distribution. At present, many details and specifics will apply to the situation in the US, but the structure of the courses and the curriculum could certainly be amended for use in the UK. What's needed here is a partner to work with NMPAN and Zarling. He sure is up for cross Atlantic cooperation.

That still leaves one crucial element: market access. Zarling believes farmers need to cooperate more and better. It could be forming a cooperative or association or just a lose network. The purpose could simply be to aggregating animals for slaughter or as sophisticated as creating a brand and coordinating marketing and sales. Nick Wallace the farmer from Iowa who is building up the '99 Counties' sales and distribution channel for meat from grass-fed animals sees his efforts as the beginning of a movement to change the regional food system not just in Iowa, but in the Midwest. And why shouldn't something similar be possible in the UK?

In 2023, I had the chance to portray six British craft butchers for a Sustainable Food Trust online series. All have been in business for a long time and as the regional economic and social circumstances changed over time, they adapted and branched out. Two of them, Robert Jones and Ieuan Edwards (aka 'The Welsh Butcher') could easily be the kind of key figures needed to create a profitable food chain for sustainable meat that links groups of farmers raising animals on grass with consumers. The links in that chain would be humane slaughter, local processing, product branding, marketing and creating market access.

Robert Jones is butcher and retail director at Walter Smith Fine Foods Ltd. He joined the company as a teenager, became an apprentice in the in-house training programme, and by the age of 27 he was the manager of a shop. "For a long time, the West Midlands had the highest

Robert Jones takes meat to where the customers are. Walter Smith meat products are now also sold at shops in garden centres.

Only few traditional butcher shops survive on Britain's High Streets.

concentration of butcher shops in the UK because of the heavy industry", he told me, there were 40 Walter Smith outlets, many within easy reach of factories in and around Birmingham, the company had a freezer centre, a pie factory, and a factory to cure bacon and prepare cuts. With the decline of the heavy industry the number of customers declined, too, many shops had to close because staff and running costs were just too high. "We need to be where the customers are", was Jones' conclusion. Of the 12 outlets Walter Smith Fine Foods operates today, only two are traditional high street butcher shops, the others are integrated into garden centres. "Garden centres are open seven days a week, they have a big car park, and the customers have disposable income", says Jones. All 10 garden centres with Walter Smith butcher outlets also have a partnership with Waitrose[136] which supplies everything except meat products. What if a group of farmers were able to offer a consistent supply of beef and pork that could be marketed as grass-fed/outdoor reared and regionally sourced or carry a brand under which the farmer group, co-op or association sold? It would be a high value addition to the range on sale at the garden centre. It would not prevent them from direct marketing their meat through farm shops or online, but help to pool resources. What if there were a designated person to deal with supply, storage, sales, packing and distribution rather than each farmer trying to do all of that in addition to their existing workload? With a steady supply and demand and such a structure in place options for on-farm slaughter or setting up slaughter modules such as those developed by Friesla or Kometos could become viable[137].

Or take Ieuan Edwards, 'The Welsh Butcher'. He was 20 when he bought his first butcher shop, his shop in Cowy recently got the Butcher Shop of the Year award. "It has its own identity", says Edwards, "we still carcass butcher here, everything we sell has been produced on the premises". Products for the 'Welsh Butcher' brand are produced in a facility at a nearby business park. Edwards told me why he expanded. "In 2000, I noticed that 'regional food' suddenly became a topic with supermarkets and started planning", says Edwards. "First, we had to choose a product. Our sausages had not only won awards, sausages are also eaten by everyone and they can be part of breakfast, lunch or dinner and they are great for a barbeque". Supermarkets don't buy from a butcher owning one shop, they need a supplier with a production facility. Edwards approached Conwy Council and was offered land in a newly opened business park. The Welsh Development Agency gave him a grant, and together with a bank loan of one million £, he bought four units in the business park. One to house the new processing facility and three to rent out and make the whole project financially viable. In the first year, ASDA (the UK arm of Walmart) came on board and stocked Edwards' sausages in four shops. In the second year, the big UK supermarket chains Tesco, Sainsbury's and Morrisons began selling his produce across shops in North Wales, Edwards broke even. He now supplies to 800 outlets and the brand is available through Ocado, an online food retailer, too. "I started out 40 years ago with no staff and £35,000 turnover a year. Today we have a turnover of £15 million and 80 to 85 full time and part time staff".

Edwards uses only Welsh beef, carrying the PGI – protected geographical indication – logo. Leeks and honey, too, are sourced from Wales. The pork shoulder meat needed for the sausages is British, farm assured and carries the UK Red Tractor label. Would there be opportunities for a cooperation between Edwards and Welsh farmers that goes beyond the PGI label? Definitely not with the goal of vertical integration, but by finding options for farmers to get animals slaughtered as humanely and close to the farm as possible while creating more consistent, efficient and profitable ways to market and sell the meat they produce.

Ieuan Edwards, the
'Welsh Butcher'.

Sausages, burger patties and
bacon are some of the products
made in the processing unit
on the outskirts of Conwy.

Public procurement could make a huge difference. At the
2024 ORFC, several sessions addressed the topic. There were
encouraging examples of cooperation between local authorities,
suppliers and famers, in some cases facilitated by NGOs working
to promote healthy food in schools, hospitals and canteens. What
can work really well in small communities becomes very difficult
in big cities. Sarah Pullen is the head of the Food System Team
at Birmingham City Council. She explained the complexity of the
system: the procurement team is tasked to source everything,
from paper clips and high viz vests to contracts for suppliers that
schools can use – the council does not buy food, it just deals with
contractors. The contracts often run over several years, budget cuts
leave no room to manoeuvre...Pullen says for Birmingham to have a
'Food System Team' is a major step in the right direction but change
will take time.

In the early 2000s, the Hospital Food Project[138] looked at options
for sustainable food procurement at London hospitals in a year-
long study. A recent initiative in California shows what scale such
a project might have. As Civil Eats reports: "In 2022, the University
of California (U.C.) system - a network of 10 campuses and five
medical centers - committed to supporting regenerative farming as
part of U.C. President Michael Drake's vision to mitigate the effects
of climate change and drive a more equitable food system.[139]"

PGI stands for 'Protected Geographic Indication'
– you know where your meat comes from.

Throughout the academic year, the campuses provide more than 600,000 meals a day. "By ensuring reliable demand for regeneratively raised meat, proponents of the system's new procurement pledge see the sizable volume giving the state's independent ranchers and rural economy a huge boost, and bolstering the local and regional meat supply chain". Santana Diaz, the executive chef at the University of California Davis Medical Center, has taken the lead in establishing a supply chain in a 250-mile radius. He tells Civil Eats: "Institutions are the perfect outlet, says Diaz, for ground, braising, and stewing meat and the other lower-value, secondary cuts that make up nearly two-thirds of every beef carcass. So featuring hamburgers, boneless short ribs, and carne asada as part of a local farm-to-fork menu offers nearby ranchers a prime bread-and-butter opportunity, he says—all the while exposing a captive audience to the value of beef raised on regenerative pasture." Most organic and regenerative farmers direct market the meat they produce, but customers usually want steaks and other prime cuts. This initiative could provide the much needed market demand for lower value cuts and give farmers and ranchers planning security. The California Department of Food and Agriculture (CDFA) promotes regenerative agriculture practices as climate friendly and the California government is channeling $600 million into the niche meat supply chain through the Community Economic Resilience Fund (CERF). The money is to secure a network of small abattoirs, processing and storage facilities.

Linking grass-based livestock rearing, humane slaughter, local processing, product branding, marketing and market access is essential for the profitability of farms which, in turn, has a huge positive impact on rural communities as a whole. "Today, livestock grazing systems are considered as the best representatives of multifunctional agriculture in Europe, as they have multiple non-market functions such as the preservation of agricultural landscapes, their contribution to rural development, including tourism and recreation, and the preservation of cultural heritage values, such as the production of traditional food products", says Alberto Bernunés[140] from the Centre for Agrifood Research and Technology (CITA) in Spain.

Nutritious and healthy meat and dairy, high animal welfare standards, humane slaughter, benefitting rural communities and the environment – what's not to like? Why isn't it happening everywhere? Animal welfare and humane slaughter will never come cheap. Meat produced to such standards, possibly certified organic or with a verifiable 'grass-fed' label will be expensive. If farmers were to calculate the true costs of slaughter, processing, packing and delivery and add a profit margin they would only be able to sell selected prime cuts into the upper end of a niche market. If they try to compete with industrial agriculture and the cheap meat from grain fattened animals raised in confinement they will likely go out of business very quickly. Yes, there is a cost of living crisis, but we also need true cost accounting[141]. Farmers cannot farm and be asked to bear the cost of sustainability, high animal welfare and humane slaughter. Which is why we have to change the narrative: the primary reason to keep livestock on grass is their function as 'land managers', maintaining habitats and providing a host of other benefits. The fact that they might provide milk during their lifetime and meat and leather after slaughter – these are secondary benefits. In my opinion, the environmental benefits of livestock farming cannot and should not be financed through the price consumers pay for dairy products and meat. We as a society have to pay for environmental services provided and that means investing tax payers' money into regenerative livestock farming and agroecology.

Chapter 13

In a time of change in agriculture, the narrative needs to change

Cows[142] (and other ruminants) are at the centre of this story. Grasslands are one of the two largest ecosystems on the planet and they are under threat. The veterinarian and author Anita Idel lays out the impact industrial agriculture has on grassland and how we got to where we are today: "As breeding focused on high yields, permanent grasslands were neglected by research, in teaching and in practice. A decline in grazing, made worse by a decrease in quality of permanent grasslands increased the usage pressure. The remaining grassland is increasingly used for silage making – mostly from a species poor seed mix with a high percentage of short rooting pasture grasses. Fertiliser is applied from slurry tankers with its axle loads contributing massively to soil compaction and thus to a reduced water absorption and storage capacity of the soil - at the expense of the quality and resilience of the permanent grassland[143]".

Today, the discussion about livestock numbers focuses on methane emissions. Feed additives such as seaweed are tested for their potential to reduce the amount of methane emitted through burping. Sheep and cattle have to stand in insulated chambers where exact amounts can be measured. Breeding sheep and cattle that naturally burp less has become a goal. Idel argues that species rich grassland such as prairies "reduce the formation of methane in rumen and in cow pads. The protection of environment and resources as well as animal and human health not only necessitate a drastic reduction in meat consumption that relies on arable land but also a massive extension of permanent or temporary grasslands and grazing. Instead, identifying ruminants as climate killers distracts from seeing fossil fuels as drivers of global growth problems. In fact, in particular the energy intensive production of synthetic chemical fertiliser with fossil methane contributes massively to the climate and biodiversity crisis. This distraction from the potential of sustainable grazing is so successful that the Irish Republic has announced plans to slaughter 200,000 cows because of the climate. (...) By ignoring the potential (of grasslands through grazing) a continued loss of large herbivores will risks to further destabilise the carbon content in soils. What's needed is more permanent grassland with grazing animals where presently cultivated slopes and meadows increasingly are eroded by heavy rainfall events".

Britain has committed to net zero by 2050 and to reach this goal we can ill afford to neglect the potential of sustainably grazed grassland. The three tier Environmental Land Management Scheme (ELMS) replaces the EU Basic Payment Scheme (BPS) and "will pay for land-based environment and climate goods and services[144]". While details for the third tier, the Landscape Recovery Scheme, are still scarce, farmers have been able to sign up for the first tier, the Sustainable Farming Initiative, SFI, since 2023. When the programme was first introduced after Britain left the EU in 2020, it was solely designed to pay farmers in England[145] 'public money for public goods' while food production was not mentioned. At the end of February 2024, the government website says: "SFI offers payments to farmers to carry out farming activities in a more environmentally sustainable way so that they can produce environmental goods and services alongside food. (...) Many of the actions we will pay for in SFI will help farmers reduce their costs and improve their efficiency as well as help improve the natural environment and reduce carbon and other greenhouse gas emissions"[146].

By February 2023, only 2,000 farmers had enrolled, a year on their number increased to 10,000 - out of roughly 100,000 farmers[147]. In autumn of last year, DEFRA published a 160 pages SFI Handbook that told farmers what measures would result in which payment. Since then, a host of measures have been added and the payment structure has changed. In regard to grassland, a positive development is that farmers now receive a much larger sum for 'maintaining species rich grassland' - the initial sum had been £182 per hectare and year, which has now been raised to £646. But as Abi Kay, the deputy editor of Farmers Weekly explained on the BBC's Farming Today programme: many schemes for upland farmers have not seen an increase in payment. By the end of this year, BPS will be cut by 50%. Upland and hill farmers who are most likely to depend wholly on livestock, "will really struggle". The problem here is that payments for measures under SFI are not additional farm income but have to replace BPS payments which have been gradually reduced since 2021 and will end in 2027.

For hill farmers, up to 95% of farm income previously came from the BPS, on upland farms where the climate is not quite so rough and pasture quality is better, direct payments made up between 30% and 60% of farm income.

In summer of 2023, I visited a number of upland farms in Northumberland to find out how farmers are coping with the reduction of BPS and how much they believe they can replace through SFI.

Most lamb outside and only have to be brought to the farm for shearing.

Graham Robson and his family farm 500 acres, they have 60 suckler cows, a Limousin cross bull, and 400 hardy northern Blackface breeding ewes. Up to the time of my visit the Robsons had not signed up for any stewardship schemes because of the management restrictions that would entail. They recently hired a consultant to help them work out which SFI schemes might provide an extra income without having to change practices on the farm too much. The 'low input grassland' scheme, and 'providing habitat for ground nesting birds' might be options. The consultant believes that, all going well, the payments for those two schemes could make up for two thirds of the BPS money.

Michael Walton farms 485 acres, has an 80 head suckler cow herd and 700 Swaledale sheep. Some of the land is in a mid-tier Countryside Stewardship scheme, which earns him £14,000. Under the EU Basic Payment Scheme he received £55,000, which made up 50 - 60% of the farm's income. From the SFI options available he will receive £4,000 at best. Which means that by 2027 her will be £50,000 worse off than pre-Brexit.

"Basic payment was free money, SFI money has strings attached", says farmer David Stanners. Farmers not only will have to find out whether they qualify for particular schemes, they also need to decide whether the restrictions that may apply to a particular scheme don't cut into profits and leave them worse off. He and his wife farm 600 acres. The 75 suckler cows are Luing[148] cattle, a crossbreed of Beef Shorthorn with Highland cattle, hardy and with excellent meat quality and there are 600 sheep. If the whole farm were to be placed into a higher tier SFI scheme, they would likely receive £36,000 - which would leave a financial hole of £30,000. And to be eligible they might have to reduce stocking numbers. The free consultancy points only to one possible new income stream: upland woodland planting on pastures. The Stanners like the idea, but there is a risk that they will not be allowed to join the scheme. The pastures are a curlew nesting area and protected. Curlews are ground breeders and don't like trees in the vicinity, which means it is up to Natural England, a nature conservation agency, to accept or deny entry into the programme.

For Nigel Moore basic payments accounted for 50 to 60% of farm income. Much of the land he farms has been in the Higher Countryside Stewardship programme for two decades. He, too, used the government's offer to get a free consultation from an agronomist. The advice was to enter additional land into a higher tier SFI scheme. But that would come with restrictions: he would have to keep a 70/30 ratio of cattle to sheep on the pastures.

He might be paid £10,000 through SFI, but would have to take 200 sheep off the pastures and introduce more cattle. It's almost impossible to figure out whether the changes would result in a net financial gain and be worth doing. "You are forced to make choices, but you don't know what the impact will be", says Moore.

There are no real options for diversification: permissions to build holiday flats or even dedicate land as a campsite are hard to get. And everything, even something as simple as a mobile chicken coop, would need investment of money the farmers don't have. In the area I visited, the local butcher who custom slaughtered animals, has closed down.

Since the time of my visit new measures have been added to SFI, some payment parameters were changed. The changes may make a slight difference to what individual farmers can expect, but the do not change the overall picture.

If the government were to truly make use of the climate potential of *grazed* grasslands, SFI would need to fund *sustainable livestock farming*.

Here are some suggestions: Financial incentives for well-managed grazing – training, practice and grants for equipment such as electric fencing. Incentives for keeping dairy cows out on grass for longer or ideally year-round. Grants for mobile milking parlours. The environmental benefits here would be particularly great because they reduce or eliminate the need for (imported) feed grain which would have been produced with fossil fuel based inputs.

These hills are made for rearing cattle and sheep. On-farm slaughter would guarantee animal welfare and be profitable for the farmers.

There should be financial incentives for integrating livestock on arable farms – to graze Herbal leys and cover crops which don't have to be terminated with herbicides such as glyphosate. Cattle spread fertiliser while grazing and they help reduce food waste when gleaning fields after harvest. The benefits of an agroecological practice[149] such as integrating livestock on mixed farms stack up and have positive knock-on effects. Implementing them should be financially rewarded because of the benefits they deliver for the *environment*. Take the practice of running chicken in a pasture once the cows have moved. By foraging for slugs and larvae they reduce the probability of future worm infections in cattle which therefore do not need to be dewormed as frequently or not at all which in turn helps to establish dung beetles and increase their numbers over time which benefits soil health...

Invest in breeding, not for yield, but hardy, dual purpose animals, well adapted to the local climate and weather. According to a report on BBC's Farming Today[150], the government in Northern Ireland is incentivising farmers to get beef animals to their finishing weight in 26 months rather than 30 months by 2027 to reduce GHGs emissions. A target that is unlikely to be reached without supplementing their diet with grains. John McLenaghan from the Ulster Farmers Union said "more product out for less product in" was a measure for more efficient farming and genetic programmes to breed animals that convert feed faster and better were being looked at. Considering the benefits of grazed grasslands, such breeding goals are steps in exactly the wrong direction.

The environmental impact study for White Oak Pastures in Georgia mentioned in the first part of this book showed that raising cattle on grass in a well managed system can have a negative carbon footprint. It's an amazing result and points at the direction research should take: Research into grass management - rotational grazing, mob grazing, adaptive multi paddock grazing – which is the best system for which geographical and climatic region? And research into cattle - which breed does well where – and programmes for developing dual purpose breeds that thrive on grass, efficient at producing milk or meat. Such research can't happen in a lab, farm trials are crucial and farmers should be financially rewarded for conducting them.

A transition to net zero cannot happen unless it is properly financed - that was the argument professor Tim Benton from Chatham House made recently on Farming Today[151]. "Change has to happen but change needs to have farmers on board and farmers need to be able to see that they will make just as much of a livelihood or even better, more of a livelihood through the process of change than being told you've got to bear all of its costs. (...) The margins that farmers can make from the market themselves have been largely squeezed and squeezed and squeezed. And they're having to bear a lot of the costs of the change, so it's kind of a double whammy and they are not getting enough money to incentivise them". Benton called for an integrated strategy to keep farmers in business and farming in a sustainable way: "Part of the answer is not just in regulations and it's not just in subsidies, it's also in getting the market to reward farmers to do the right thing. We need an integrated strategy and we need to make sure that there is money behind it either through changing food prices which again leads to other social issues or by ensuring that there is public money for public goods in a way that rewards farmers to do the right thing".

As Benton says: the money needed cannot solely come from an increase in food prices. "It is a generally accepted maxim of governments when facing environmental issues that 'the polluter should pay'. By the same logic society should be prepared to reward the 'anti-polluters': those who are actively engaged in the conservation of wildlife, management of water and sequestration of carbon since they offset the anthropogenic insults of the rest of us. This can be best achieved via taxation through redirection of agricultural subsidies"[152], says John Webster, professor emeritus and founder of the UK Farm Animal Welfare Council.

Maybe additional taxation is necessary. The German government has recently introduced a five category mandatory animal welfare labelling scheme[153]. Category 1 meat comes from animals in confinement industrial agriculture, meat from animals in a certified organic system is labelled category 5. At present, only pork has to carry the new label, standards for meat from beef and poultry will be introduced successively. The aim is to have all producers upgrade their systems to a standard of category 3 and above which provide the animals with more space and access to the outdoors. Upgrading barns is expensive and farmers will need financial support or they will go out of business. Different ways how better animal welfare could be financed are under discussion. Among the ideas are an "animal welfare cent": animal welfare organisations would like an extra Euro 0.40 per kilogram of meat. Another option might be an increase of VAT from 7% to 19% for meat products. I think there is room to be creative. How about taxing all meat and dairy products from animals *not* in a grass-based system, with the proceeds going to livestock farmers with sustainably managed pastures?

It is amazing what policies politicians can come up with once they take the climate crisis and questions of food security seriously. Piet Adema became minister for agriculture in the Netherlands in 2022. In a recent interview he said cattle numbers would have to be reduced by 20%. "Most farmers I talked to can accept that as long as they can still make a living[154]". The Dutch government has developed a scheme that offers a buy-out to livestock farms with particularly high GHG emissions if the farmers commit to not farm in a similar manner in a different location. "Almost 1,300 livestock farmers have applied for the scheme. We are in the middle of checking eligibilities. Up to now, 50 farmers have signed a contract". The Dutch government also offers large financial incentives for farmers to switch from high input industrial systems to organic farming, but lack of demand for certified organic produce is a major problem, says Adema. A 50 million Euro information campaign will aim to educate consumers about organic farming.

"We have to increase the percentage of organic agriculture. It is a major element for more sustainability in agriculture".

The Netherlands, a country the 6th of the size of the UK, are the second largest exporter of agricultural products[155] in the world and the 4th largest dairy producer[156]. And yet it seems possible to change the narrative and switch the focus to sustainability and agroecology.

Cows and the eco-services they provide place them at the centre of the story. Sustainable food production is not possible without livestock. Small abattoirs, MSUs and slaughter boxes are necessary parts of the infrastructure. Well-trained butchers, skilled processors and creative chefs are essential, but livestock farmers make or break the system.

Two of the farmers I talked to, Mechthild Knösel in Germany, and Jeff Rogers in the US, have thought long and hard about keeping livestock and the ethics of meat production. To them, humane slaughter, an animal dying with dignity and without fear or stress, is part of their obligation towards the animals in their care. In her early morning, very private ceremony, Mechthild Knösel, thanks the animals for what they have given to the farm. We all need to consider the central role of livestock, cows in particular, in food production, for farm income, in rural economies and for the environment.

Acknowledgement

My thanks and gratitude to all the people who have given me their time, shown me around their farms, let me witness what happens in small abattoirs and MSUs, explained how meat is processed and made into fantastic high value products. These farmers, ranchers, butchers and slaughtermen, processors, chefs and educators are true pioneers, passionate about creating a food system that prioritises animal welfare and the environment while also allowing everyone involved to operate profitably and earning at least a living wage.

I hope this book will help to convince others that to preserve grasslands we need ruminants, in particular cattle. The co-evolution of grasslands and ruminants that graze them created the soils that make it possible for us to grow food. For that alone livestock deserve a good life and a good death.

My thanks go to the Sustainable Food Trust, in particular to Megan Perry who was up for this project from the start and to Alicia Miller for doing the first edit. David Zarling from NMPAN, not only shared his insights but introduced me to a number of people in the niche meat sector who are key in building a local, sustainable and profitable meat supply chain in the Pacific Northwest. Anita Idel patiently explained why grasses are different, what potential they hold and why grasslands need to be grazed by bovines.

And I am beyond grateful to Martin Kunz, my husband, who took thousands of pictures, drove thousands of miles in Germany and the US and stood by me through the emotional rollercoaster ride this project turned out to be. Without his support and love there would be no book.

Endnotes

[1] Gabe Brown: Dirt to Soil, Chelsea Green, 2018, page 11

[2] loc. cit. page 22

[3] Nicole Masters: For the love of Soil, Printable Reality, 2019, page 28

[4] Marianne Landzettel: Regenerative Agriculture - Farming with Benefits, WUS, 2021, page 1

[5] loc. cit. page 45ff

[6] loc. cit. page 25

[7] loc. cit. page 31

[8] loc. cit. page 43

[9] loc. cit. page 216

[10] loc. cit. page 215

[11] https://www.advancingecoag.com/plant-health-pyramid

[12] loc. cit. page 220

[13] https://www.csuchico.edu/regenerativeagriculture/ra101-section/ra101-definitions.shtml

[14] https://www.csuchico.edu/regenerativeagriculture/ra101-section/index.shtml

[15] https://attra.ncat.org/publication/integrating-livestock-and-crops-improving-soil-solving-problems-increasing-income/

[16] Nic Lampkin, Mark Measures, Susanne Padel: organic Farm Management Handbook 2023, Organic Research Centre, 2023

[17] Farmers Weekly: Herbal leys and bicrops boost wheat premium, 15 December 2023, page 90

18 https://www.csuchico.edu/regenerativeagriculture/ra101-section/integrating-livestock.shtml

19 SOM Soil Organic Matter

20 https://live-the-pasture-project.pantheonsite.io/wp-content/uploads/2020/02/Grazing-Cover-Crops-How-To-Guide.pdf

21 Richard Young in: Joyce D'Silva, John Webster, Eds.: The Meat Crisis, Routledge, 2nd edition, 2017, page 99

22 https://www.martin-haeusling.eu/images/publikationen/Klimawandel2020_EnglischeVersion_final.pdf, page 37

23 Anita Idel: Koevolution von Grasland und Weidetieren Potenziale nachhaltiger Beweidung für Bodenfruchtbarkeit, Klimaentlastung, und biologische Vielfalt, Kritischer Agrarbericht 2024 (translation by ML)

24 loc. cit. page 42

25 Colin Tudge in: Joyce D'Silva, John Webster, Eds.: The Meat Crisis, Routledge, 2nd edition, 2017, page 13

26 Roger Shinn, Lunne Pledger: Grass-Fed Beef for a Post-Pandemic World, Chelsea Green, 2022, page 52

27 https://blog.whiteoakpastures.com/hubfs/WOP-LCA-Quantis-2019.pdf?hsCtaTracking=6d515b16-e2ed-4bea-a286-a7433c983b81%7C7a0781f6-8e32-4e28-89e9-563565ab2eea

28 https://www.greenpeace.de/publikationen/Milchk%C3%BChe%20in%20Deutschland.pdfhttp

29 https://sustainablefoodtrust.org/news-views/grass-fed-dairy/

30 loc. cit. page 13

31 Joyce D'Silva in: Joyce D'Silva, John Webster, Eds.: The Meat Crisis, Routledge, 2nd edition, 2017, page 53ff

32 loc. cit. page 39

33 loc. cit. page 39

34 Tara Garnett: in: Joyce D'Silva, John Webster, Eds.: The Meat Crisis, Routledge, 2nd edition, 2017, page 31

35 loc. cit. page 14ff

36 https://defrafarming.blog.gov.uk/graze-with-livestock-to-maintain-and-improve-habitats/

37 https://www.forestryengland.uk/article/managing-deer-the-nations-forests

38 Quote from Lindsay Whistance during the Livestock in the Landscape session at the ORFC on January 4th, 2024

39 https://www.vegansociety.com/go-vegan/definition-veganism

40 Yellowstone series 4, episode 5, 22'25

41 https://asmith.ucdavis.edu/news/bees-per-almond

42 https://usafacts.org/articles/what-is-the-loss-of-bees-costing-the-us/

43 https://www.earthday.org/you-are-what-you-eat-plastics-in-our-food/

44 Chris Smaje: Saying No To A Farm-Free Future, Chelsea Green Publishing 2023

45 transcript from ORFC recording https://www.youtube.com/watch?v=FuvTcC-QQUQ

46 https://sustainablefoodtrust.org/wp-content/uploads/2022/03/Re-localising-farm-animal-slaughter-low-res.pdf

47 https://www.sheffield.ac.uk/news/local-food-suppliers-proved-their-value-during-pandemic-so-how-do-we-ensure-they-thrive

48 loc. cit. page 13

49 Email communication with Amy Quirk

50 https://www.mapize.com/map/bovine-abattoir-services/

51 https://longreads.com/2024/02/15/rise-of-women-butchers/

52 https://apgaw.org/2020/06/07/apgaw-publishes-report-on-small-abattoirs/

53 https://sustainablefoodtrust.org/news-views/abattoir-sector-group-launched-to-promote-a-thriving-uk-network-of-sustainable-and-local-abattoirs/

54 Interview aired on Radio 4, Farming Today, on August 31st, 2022. Transcript by the author.

55 https://sustainablefoodtrust.org/wp-content/uploads/2023/05/Final_abattoir_users_survey_web.pdf

56 https://sustainablefoodtrust.org/news-views/progress-for-local-abattoirs-at-orfc/

57 https://sustainablefoodtrust.org/news-views/progress-for-local-abattoirs-at-orfc/

58 BBC Farming Today, 16.2.2024

59 The alternative is to let them starve – as happens in India. Under the title "Suffering herds", the Economist Feb. 3rd, 2024, page 49) described the effects of a slaughter ban: "The BJP government of the big northern state of Uttar Pradesh (UP), home to an estimated 240m people and perhaps 20m cattle, is conducting a new bovine census. The point of this, says the government, is better cow protection. The effort highlights a glaring

cow-related contradiction in the BJP's Hindu nationalist ideology. The party says it wants to protect cows, which are associated with divine beneficence and venerated by Hindus. Yet its pro-cow policies, including bans on cow slaughter, appear to be detrimental to cattle welfare. They are thought to be causing an increase in stray cows, typically male calves and aged milkers which, having little commercial value, are let loose by their owners. Abandoned, they feed on plastic bags and other rubbish, cause car crashes and raid farmers' crops. There were an estimated 5m stray cattle in India in 2019, including 1.2m in up. The government reckons the number has since increased."

60 Tara Garnett: loc. cit. page 37

61 loc. cit. page 44

62 John Webster: Joyce D'Silva, John Webster, Eds.: The Meat Crisis, Routledge, 2nd edition, 2017, page 135

63 loc. cit. page 121

64 loc. cit. page 120

65 loc. cit. page 126

66 https://londoncowgirl.com/

67 with an M.A. in psychology, University of Braunschweig, Germany and a Diploma in Behavioural Psychotherapy, Middlesex Hospital Medical School, UCL, London

68 https://www.wb6cif.eu/wp-content/uploads/2021/08/CELEX_32021R1374_EN_TXT-1.pdf page 7

'CHAPTER VIa: SLAUGHTER AT THE HOLDING OF PROVENANCE OF DOMESTIC BOVINE, OTHERSTHAN BISONS, AND PORCINE ANIMALS AND DOMESTIC SOLIPEDS OTHER THAN EMERGENCY SLAUGHTER
Up to three domestic bovine, others than bisons, or up to six domestic porcine animals or up to three domestic solipeds may be slaughtered at the same occasion at the holding of provenance, when authorised by the competent authority in accordance with the following requirements:
(a) the animals cannot be transported to the slaughterhouse, to avoid any risk for the handler and to prevent any injuries to the animals during transport;
(b) there is an agreement between the slaughterhouse and the owner of the animal intended for slaughter; the owner must inform the competent authority in writing of such an agreement;
(c) the slaughterhouse or the owner of the animals intended for slaughter must inform the official veterinarian at least three days in advance of the date and time of intended slaughter of the animals;
(d) the official veterinarian who carries out the ante-mortem inspection of the animal intended for slaughter must be present at the time of slaughter;
(e) the mobile unit to be used for the bleeding and transport of the

slaughtered animals to the slaughterhouse must allow their hygienic handling and bleeding, and the proper disposal of their blood and must be part of a slaughterhouse approved by the competent authority in accordance with Article 4(2); however the competent authority may allow bleeding outside the mobile unit if the blood is not intended for human consumption and the slaughter does not take place in restricted zones as defined in Article 4(41) of Regulation (EU) 2016/429 of the European Parliament and of the Council * or establishments in which animal health restrictions are applied in accordance with Regulation (EU) 2016/429 and any acts adopted on its basis;
(f) the slaughtered and bled animals must be transported directly to the slaughterhouse hygienically and without undue delay; removal of the stomach and intestines, but no other dressing, may take place on the spot, under the supervision of the official veterinarian; any viscera removed must accompany the slaughtered animal to the slaughterhouse and be identified as belonging to each individual animal;
(g) if more than two hours elapse between the time of slaughter of the first animal and the time of arrival at the slaughterhouse of the slaughtered animals, the slaughtered animals must be refrigerated; where climatic conditions so permit, active chilling is not necessary;
(h) the owner of the animal must inform the slaughterhouse in advance of the intended time of arrival of the slaughtered animals, which must be handled without undue delay after arrival at the slaughterhouse;
(i) in addition to the food chain information to be submitted in accordance with Section III of Annex II to this Regulation, the official certificate set out in Chapter 3 of Annex IV to Implementing Regulation (EU) 2020/2235 must accompany the slaughtered animals to the slaughterhouse or be sent in advance in any format

69 https://uria.de/mobile-schlachtbox/

70 A dried off cow has finished her lactation cycle, she will not give milk until she gives birth to another calf.

71 In an abattoir, animals are rendered unconscious through stunning – with electricity as with these pigs, through a shot from a bolt gun or with gas, but they die through bleeding out.

72 https://mlr.baden-wuerttemberg.de/de/unser-service/presse-und-oeffentlichkeitsarbeit/pressemitteilung/pid/antragsverfahren-zur-foerderung-von-regionalen-schlachthoefen-nach-tierwohl-kriterien-ist-geoeffnet/

73 https://albert-schweitzer-stiftung.de/aktuell/kleine-schlachthoefe-fehlbetaeubungen

74 Unabhängige Bauernstimme, September 2020, page 3

75 https://ahdb.org.uk/knowledge-library/how-pre-slaughter-stress-impacts-meat-quality

76 https://edoc.ub.uni-muenchen.de/23922/1/Wullinger-Reber_Hanna.pdf

77 Between November 2017 and March 2017 Wullinger-Reber was present on 11 slaughter days at the Lindner Group farm and analysed 66 pigs.

78 Nicolette Hahn Niman: Defending Beef, 2nd edition, Chelsea Green, 2021

79 David R. Montgomery, Anne Biklé: What Your Food Ate, W.W. Norton & Company, 2022

80 loc.cit. page 195

81 Conjugated linoleic acids (CLA). According to Montgomery and Anne Biklé CLA from dairy and meat provides a wide range of health benefits: it has been found to help prevent certain types of cancer, type 2 diabetes and plaque building in arteries. See: What Your Food Ate p. 290 ff

82 Fred Prvenza: Nourishment, Chelsea Green, 2018, page 33

83 loc. cit. page 180

84 loc. cit. page 135

85 Shinn, Pledger: loc. cit. page 63ff

86 loc. cit. page 5

87 https://www.sheffield.ac.uk/news/nr/land-cover-atlas-uk-1.744440

88 https://modernfarmer.com/2024/01/meet-the-ranchers-trying-to-restore-grasslands/

89 Anita Idel cit. loc. Kritischer Agrarbericht

90 https://www.martin-haeusling.eu/images/publikationen/Klimawandel2020_EnglischeVersion_final.pdf

91 loc.cit. page 37

92 https://www.pbs.org/kenburns/the-american-buffalo

93 https://grist.org/beacon/bringing-bison-back-to-tribal-lands/

94 https://www.usda.gov/media/press-releases/2023/10/12/usda-invests-bison-purchase-pilot-incorporating-indian-country

95 https://ambrook.com/research/supply-chain/food-sovereignty-USDA-bison

96 According to their website: Ambrook Research is editorially independent but backed by Ambrook, a company making sustainability profitable in natural resource industries, starting by providing back-office financial tools for farmers.

97 Ridge Shinn, Lynne Pledger, loc. cit. page 137

98 Timothy Pachirat: Every Twelve Seconds, Yale University Press, 2011, page 23

99 Timothy Pachirat, loc. cit. page 145

100 Timothy Pachirat, loc. cit. page 149

101 Timothy Pachirat, loc. cit. page 148

102 Timothy Pachirat, loc. cit. page 54

103 Timothy Pachirat, loc. cit. page 158

104 RSPCA, Royal Society for the Prevention of Cruelty to Animals, British animal welfare charity

105 https://kb.rspca.org.au/knowledge-base/what-does-the-term-humane-killing-or-humane-slaughter-mean/

106 https://www.fsis.usda.gov/sites/default/files/media_file/2020-08/93-016F_0.pdf

107 https://www.nichemeatprocessing.org/hazard-analysis-critical-control-point-haccp/

108 https://www.whitehouse.gov/briefing-room/statements-releases/2022/01/03/fact-sheet-the-biden-harris-action-plan-for-a-fairer-more-competitive-and-more-resilient-meat-and-poultry-supply-chain/

109 https://www.centerforfoodsafety.org/press-releases/6729/statement-from-center-for-food-safety-and-food-and-water-watch-regarding-usdas-dangerous-slaughterhouse-self-inspection-program

110 https://www.pbs.org/wgbh/pages/frontline/shows/meat/evaluating/haccp.html

111 loc.cit, page 95ff

112 We meet up with Jacob Fortney after he finished work for the day. He wants to expand his butcher business and wants to show us his equipment. As independent butcher he hopes to provide a bespoke on farm service. In the US, animals can be slaughtered on farm, and neither the animal nor the meat have to be inspected, as long as both meat and meat products are only consumed by the farmer and his family. This is called 'custom exempt slaughter'. Because he has no cooler, Fortney is only allowed to slaughter from October to May. To start his business, he has bought about $6,000 worth of equipment, the trailer belongs to his Dad. It's his second year as custom butcher. In the 2021/2022 season he had 11 bookings. At the start of this year's season he has already done two. He charges $1.10 per pound hanging weight and 5¢ per mile, the farms have to be within a day's drive. He stuns and bleeds the animal out, skins and guts it and then breaks it down into about 12 parts so that he is able to lift them by himself. He places the meat into buckets on the trailer and delivers them to the plant in Bowdon. On Sunday he processes it into minced meat or burger patties ready for pick-up. BMP charges another ¢50 per pound for the use of the grinder and packaging.

113 https://www.nd.gov/ndda/news/farm-school-efforts-ramp-north-dakota

114 Sarah Vogel: The Farmer's Lawyer Bloomsbury, 2021 page 13

115 https://friesla.com/

116 Next door to the facility is the Washington State University BreadLab, which we visited a few years earlier. https://breadlab.wsu.edu/ The facility is the brainchild of Stephen Jones, a plant geneticist and wheat breeder who is researching wheat, barley, buckwheat and other small grains to identify varieties that do well on farms in the Pacific Northwest. The BreadLab identifies which varieties are most suitable for craft baking, cooking, malting, brewing, and distilling.

117 https://podcasts.apple.com/gb/podcast/the-meat-block/id1217207010

118 https://www.foodsafetyuniversity.com/

119 https://meatprocessoracademy.com/curriculum

120 https://meatprocessoracademy.com/mastermind-program

121 https://www.whitehouse.gov/briefing-room/statements-releases/2022/01/03/fact-sheet-the-biden-harris-action-plan-for-a-fairer-more-competitive-and-more-resilient-meat-and-poultry-supply-chain/

122 https://meatingplacepdx.com/custom-cutting/ The website lists the prices and fees for slaughter, cut & wrap and processing. In the US, pigs, too can be slaughtered in an MSU because the butchers routinely skin them and no dehairing tank is needed.

123 https://rangepartners.com/

124 Kari Underly The Art of Beef Cutting. A meat profession's guide to butchering and merchandising. John Wiley & Sons, Inc. 2011

125 One of the textbooks used in the course is Kari Underly's The Art of Beef Cutting.

126 Beth Hoffman: Bet the Farm. Island Press 2021, p183ff

127 Practical Farmers of Iowa is a farmer organisation facilitating knowledge exchange and farmer led research. https://practicalfarmers.org/

128 https://99counties.com/

129 Shinn, Pledger, cit. loc. page 162

130 http://www.hnvlink.eu/download/Sweden_Mobileabattoir.pdf

131 https://kometos.com/news/kometos-oy-delivers-a-mobile-slaughterhouse-for-beef-cattle-to-france/

132 https://www.foodnavigator.com/Article/2016/09/21/France-powers-up-Sweden-s-mobile-abattoir#

133 https://www.gov.scot/binaries/content/documents/govscot/publications/research-and-analysis/2020/03/assessing-viability-sustainability-mobile-abattoirs-scotland/documents/assessing-viability-sustainability-mobile-abattoirs-scotland/assessing-viability-sustainability-mobile-abattoirs-scotland/govscot%3Adocument/assessing-viability-sustainability-mobile-abattoirs-scotland.pdf

134

135 https://ambrook.com/research/author/hannah-macready

136 Waitrose is an upmarket UK supermarket chain

137 This option would of course require the UK government to change the regulatory framework.

138 https://www.iatp.org/sites/default/files/421_2_78534.pdf

139 https://civileats.com/2024/03/04/regenerative-beef-gets-a-boost-from-california-universities/

140 Alberto Bernunés in: Joyce D'Silva, John Webster, Eds.: The Meat Crisis, Routledge, 2nd edition, 2017, page 77

141 https://sustainablefoodtrust.org/wp-content/uploads/2022/01/Website-Version-The-Hidden-Cost-of-UK-Food_compressed.pdf

142 Pigs should of course be slaughtered humanely too, on farm, or as close to it as possible. Pigs are very important for (small) diversified farms, but their role is more important in a rotational system, not so much for permanent grassland. That is the reason for focussing on ruminants here.

143 Anita Idel: Koevolution von Grasland und Weidetieren Potenziale nachhaltiger Beweidung für Bodenfruchtbarkeit, Klimaentlastung, und biologische Vielfalt. Kritischer Agrarbericht 2024 (translation by ML)

144 https://www.gov.uk/government/publications/environmental-land-management-update-how-government-will-pay-for-land-based-environment-and-climate-goods-and-services/environmental-land-management-elm-update-how-government-will-pay-for-land-based-environment-and-climate-goods-and-services

145 As devolved nations, Scotland, Wales and Northern Ireland draw up their own farm policies.

146 cit.loc.

147 BBC Farming Today, 14.2.2024

148 Luing is a small island off the west coast of Scotland, the breed is extremely hardy and the cows know how to look after their calves even in foul, windy weather. David Stanners is a member of the Luing breeding society and finds the animals are doing well on higher pastures. And preparing two steaks upon our return home, we can attest to the excellent taste and texture of the meat.

149 The California Healthy Soils Initiative is a programme by the California Department of Food and Agriculture which "funds on-farm demonstration projects to promote statewide implementation of conservation management practices that sequester carbon, reduce GHGs and improve soil health". https://www.grants.ca.gov/grants/healthy-soils-demonstration-program/

The programme is funded from the California's cap and trade proceeds. It is beyond the scope of this book to discuss carbon credits and 'carbon farming'. The approach in California shows that incentivising agroecological practices is a tried and tested approach increase carbon sequestration in agriculture.

150 BBC Farming Today, 14.2.2024

151 BBC Farming Today, 7.2.2024

152 John Webster, loc.cit. page 136

153 https://www.tierhaltungskennzeichnung.de/anwender/ueberblick

154 Interview in Tagesspiegel Hintergrund Agrar & Ernährung, 19.2.2024, (Translation by ML)

155 https://www.trade.gov/country-commercial-guides/netherlands-agriculture

156 https://oec.world/en/profile/bilateral-product/milk/reporter/nld?redirect=true

Index

A

abattoir 6, 7, 8, 10, 11, 24
34, 38–47, 49–51, 55, 58–61,
66, 69, 71–73, 75, 77–89, 91,
95–101, 104, 105, 108–110,
112–114, 117, 129, 132, 133,
135–137, 141, 143, 145, 147,
149, 163, 165, 184, 194, 198,
199–205, 211, 222

animal handling........ 6, 62, 65,
103, 154, 178, 194, 204

animal welfare 8, 11, 41,
42, 45–49, 55, 59, 89, 90, 94,
98, 99, 111, 114, 115, 129,
201, 203, 212, 218, 221

apprenticeship........ 45, 51, 52,
83, 85, 88, 101, 103, 104, 113,
183, 185, 186

arable 11, 12, 20–23, 25, 32,
33, 47, 48, 62, 90, 113, 120,
132, 138, 214, 219

ASG........................... 42

B

bacon........ 151, 155, 157, 175,
208, 210

Badlands National Park.. 1, 119,
123

beef................ 6, 10, 11, 23,
28, 29, 31, 37–40, 49, 51, 55,
59–61, 67, 71, 83, 84, 88–90,
96, 102, 109, 110, 113–117,
127, 130, 133–136, 138, 140,
141, 143, 145, 148–150, 152,
156, 160, 167, 169, 172–176,
179, 182, 183, 185, 188, 190,
192, 193, 202, 204, 208, 209,
211, 217, 220, 221

biodiversity 7, 19, 22, 24, 48

Bismarck........... 140, 142–144

bison......... 119, 120, 123–127,
136, 138, 203

bleed out 59, 60, 66, 69, 74,
75, 78, 110, 134, 146–148, 165

blood......... 59, 70, 72, 75, 110,
115, 165, 200

Bonkhoff, Klaus 70, 71, 204

Bowden........... 136, 137, 139,
142, 144, 203

Brexit................. 39, 50, 217

Brown, Gabe. 12, 14, 18, 19, 135

butcher 9, 10, 11, 40–44,
46, 49–53, 57–62, 66, 67,
69–75, 81–90, 94, 94–98,
100–106, 108–114, 117, 129,
132, 134, 136, 137, 140–142,
145–148, 150, 151, 153–156,
159, 160, 166, 168–173,

179–187, 194, 197, 198, 200, 201–210, 218, 222
butcher training......... 101–117
butchering 78, 96

C

Callicrate, Mike 132–134, 144, 199
calves......... 49, 62–64, 67, 73, 74, 78, 79, 117, 142, 202
carbon 14, 15, 19, 24, 25, 27–29, 33, 37, 48, 52, 121, 122, 124, 214, 220, 221
carcass 40, 43, 58, 66, 69–71, 74–77, 80, 94, 96, 97, 109, 110, 127, 128, 133–135, 147, 148, 155, 165, 169, 170, 179, 182, 192, 194, 198–200, 204, 209, 211
cattle.......... 10–12, 21, 23–25, 27–29, 31–34, 37, 39–41, 48, 51, 52, 55, 59–62, 65, 67, 71, 85, 87–90, 95, 99, 108–110, 113, 116, 117, 122, 126–128, 132–134, 136, 138, 139, 141, 143, 145, 146, 149, 150, 174–177, 182, 183, 188, 192, 194, 198, 203, 204, 213, 217–221
charcuterie............ 74, 75, 82, 151, 154, 157, 160, 179, 194, 201, 205
Chicago........... 126, 127, 129, 155, 156, 172, 184, 189, 190, 192, 193
Chico University 19
chute........ 128, 129, 134, 136, 176, 203
co-op 136, 137, 138, 143, 146, 149–153, 155, 159, 188, 194, 206, 208

Colorado . 27, 132, 134, 135, 190
cooler.......... 73, 98, 133, 134, 136, 137, 148, 149, 150, 164, 165, 168, 169, 174, 175
corn........... 16, 18, 19, 28, 31, 44, 51, 53, 98, 117, 120, 123, 126, 130, 138, 139, 148, 165, 169, 191
cover crops .. 19–22, 24, 25, 219
COVID .. 10, 43, 78, 99, 142, 143
cows.......... 11, 20, 21, 25, 26, 28–31, 49, 50, 62–65, 67, 68, 78, 79, 116, 191, 202, 213, 214
cure................... 77, 82, 83, 97, 111, 154, 157, 159, 161, 172, 208
Cured Processing 154, 157–158, 159, 187, 195
cutting 103, 114, 127, 136, 153, 154, 156, 168, 171, 174, 179, 181, 183, 184, 187, 198, 205

D

dairy 7, 10, 11, 28–31, 34, 47, 48, 50, 51, 62, 65, 67, 68, 73, 77–79, 85, 114, 116, 200, 202, 212, 218, 221, 222
de-hairing................. 74, 76
de-hiding.................. 79, 96
death.............. 7, 10, 11, 38, 39, 44, 47, 59, 70, 89, 98, 99, 129, 233
DEFRA........... 33, 44, 45, 214
direct market...... 41, 88, 90, 98, 135, 139, 170, 201, 203, 205, 208, 211
direct payment................216
direct sales 40, 41, 145, 153
Duck Fest....................140
Dust Bowl................51, 119

Dutton, John 35
diversification 41, 218

E

education 62, 102, 112, 130,
   183, 184
Edwards, Ieuan, The Welsh
Butcher 207, 209, 210
ELMs 40, 14, 214
environment 12, 18, 27–31,
   33, 33–36, 40, 41, 47, 48, 57,
   109, 117, 122, 173, 189, 203,
   205, 206
EU 38, 40, 45, 50, 52,
   55, 57–60, 62, 67, 84, 113, 197,
   200, 205, 214, 217
EU framework.... 58–60, 62, 200

F

farmer.... 6, 7, 10–13, 16, 18–20
   23–25, 30, 31, 38–42, 44–48,
   50–53, 55, 57–62, 65–67, 70,
   71, 73, 74, 77–80, 83–91,
   98, 104, 111, 113, 113, 117,
   122, 127, 129, 132, 135–138,
   141, 143–146, 148–156, 159,
   163, 166, 170, 173, 174, 178,
   179, 181, 182, 184, 186–189,
   192–209, 211, 212, 214
FDA................... 126, 129
fixate.. 58, 59, 62, 66, 68, 69, 70,
   128, 147
Flick, Mike 138, 139
food system....... 25, 32, 38, 41,
   113, 189, 190, 205, 207, 210
Fortney, Jacob 140, 141
Fox, Del.......... 146–148, 150
Friesla 145, 209

G

Germany 29, 30, 50, 53,
   55, 57, 58, 62, 67, 72, 73, 82,
   94, 99, 100, 100, 104, 108,
   112, 129, 194, 197, 200, 201,
   203–206, 222
grass-fed............. 28, 29, 38,
   51, 52, 68, 115–117, 140, 188,
   189, 192, 193, 195, 202, 207,
   208, 212
grasses 122, 213
grassland 12, 23, 25, 27,
   30, 33, 37, 47, 51, 52, 117, 119,
   120, 122, 124, 201, 213, 214,
   217, 218, 220
Gray, Marena 166, 168, 169, 171
grazing...... 1, 20, 22–31, 33, 34,
   37, 40, 47, 48, 51, 59, 64, 68,
   78, 79, 90, 108, 117, 120–123,
   189, 212–214, 218–220

H

HACCP...... 130, 131, 133, 145,
   154, 158, 160, 170
ham.... 74, 82, 97, 98, 111, 151,
   157, 159, 161, 172, 175, 201
Hart, Corry........ 137, 138, 143
health
- animal 13, 30, 40, 42,
   45, 49, 64, 80, 85, 107, 123
- human............... 11, 36, 47,
   115–117, 122, 131, 155, 171,
   173, 179, 189, 190, 193, 205,
   206, 210, 212, 213
- soil 7, 14, 15, 18, 19, 24,
   33, 34, 36, 117, 219
herd............... 23, 30, 41, 49,
   50, 56, 59, 60, 62–65, 67, 68,
   70, 97, 108, 113, 123, 125,
   126, 138, 187, 200, 217

Hoffman, Beth 188, 189, 192
hunters 101, 139, 140

I

Idel, Anita 25, 27, 31, 121, 122, 213, 233
Iowa 159, 188, 189, 190, 193, 207
Island Grown Farmers Cooperative 145, 150–152, 155, 159, 163, 194, 199

J

JBS 138, 139, 177
Jones, Robert 207, 208

K

Kansas 132, 135, 177
killing 49, 55, 62, 75, 76, 90, 96, 103, 112, 127, 128, 129, 200, 203
kill floor 76, 127, 128, 136, 140, 176, 178, 179, 194
Kling, Thomas 74, 75, 76, 80–85
Knösel, Mechthild 62, 64–70, 79, 95, 108, 166, 200, 202
Konrad, Gero 87, 88, 89, 90

L

lamb 29, 38, 39, 41, 88, 89, 115, 152, 153, 156, 162–167, 169, 172–175, 183, 204, 216
land management 23, 29, 32, 33, 40, 214
Legau 73, 80, 81, 84–86, 114, 194, 202, 203
Lindner Group 105, 108, 110, 111–113, 201

livestock 6, 11, 12, 14, 17, 20–25, 27, 31–34, 37, 38, 41, 44, 45, 47, 48, 52, 55, 80, 89, 90, 116, 117, 132, 135, 149, 159, 182, 185, 197, 198, 212, 213, 215, 218, 219, 22, 222
local 6, 7, 8, 38–44, 46–48, 50–53, 55, 66, 69, 73, 87, 99, 104, 105, 113, 125, 126, 132, 137, 140, 143, 150, 155, 156, 163, 171, 173, 174, 181, 189, 197, 201, 202, 204, 205, 207, 210–212, 218, 220

M

marketing 52, 143, 161, 181, 187–190, 202, 205–208, 212
Masters, Nicole ... 13, 14, 16, 17
Meat Processor Academy ... 161
- cutting 103, 114, 127, 136, 153, 154, 156, 168, 171, 174, 179, 181, 183, 184, 187, 198, 205
- meat production 41, 50, 52-53, 126, 222
- meat products 52, 85, 102, 104, 109, 170-171, 192, 206, 208, 244
- quality 42, 51, 64, 68, 77, 84, 85, 88, 94, 102, 107, 114–116, 129, 170, 180, 187, 202, 217
Meating Place 162, 166–171, 187, 204, 205
methane 28, 29, 37, 213, 214
Mettrick, John 40, 42, 204
Meyer, Ben ... 171–179, 186, 205
microorganisms 17, 130
Monbiot, George 36
Monson, Lauren 141, 142

MSU 52, 58, 67, 72, 96, 109, 110, 133, 134, 135, 44–152, 154, 163, 164, 170, 171, 194, 197–199, 204–206, 222

N

99 Counties.. 189, 192, 193, 207
net zero.............52, 214, 220
NMPAN, Niche Meat Processor Assistance Network 52, 130, 154, 160, 162, 174, 178, 186, 194, 206, 207
North Cascades Meats......149
North Dakota 12, 135, 136, 138, 139, 141–144, 203
North Dakota State University, NDSU.........................141
NWMPA, Northwest Meat Processors Association...... 52, 185, 186

O

Oregon............. 64, 145, 162, 166, 167, 178, 185, 204
Oregon State University, OSU .. 130, 170
ORFC 33, 37, 44, 45, 210
organic........ 13, 18, 19, 21–24, 30, 33, 41, 44, 55, 61, 62, 73, 84, 88, 94, 100, 105, 107, 108, 111–113, 116 156, 163, 166, 172, 173, 190, 195, 204, 211, 212, 221, 222

P

Pacific Northwest... 50, 52, 144, 150, 154, 156, 159, 163, 172, 174, 175, 185, 206

pasture............ 11, 18, 21–23, 27–31, 33, 37, 41, 48, 52, 55, 58–60, 62–64, 66, 67, 70, 73, 90, 91, 94–96, 107–110, 113, 117, 138, 159, 162, 163, 174, 180, 211, 213, 216–221
- permanent...........21–23, 27, 30, 62, 107, 138
Perry, Megan 44, 45
pesticides......... 11, 16, 17, 23, 30, 31, 35, 190
pigs 6, 11, 32, 47, 74, 76, 80, 85, 88–96, 98–100, 107, 109, 110, 112, 115, 132, 134, 136–138, 140, 141, 145, 146, 153, 169, 172–178, 192, 201, 203, 204
pork....... 10, 11, 29, 80, 85, 94, 97, 115, 130, 131, 156, 159, 167, 173, 175, 183, 185, 190, 193, 208, 209, 221
Portland....... 52, 162, 166, 167, 170, 171–173, 185, 204
processing..... 7, 50, 52, 53, 58, 61, 75, 99, 100, 111, 126, 127, 129, 130, 134, 135, 136, 137, 141, 144, 145, 151, 153–159, 161–173, 175, 177, 179, 182, 187, 193–195, 197, 201, 203, 205–207, 209, 201, 212
procurement. 104, 143, 210, 211
profitable...... 30, 47–53, 83, 98, 99, 111, 159, 162, 173, 174, 194, 196, 197, 205–207, 209, 218

Q

Quirk, Amy............. 39, 40, 41

R

Range® Meat Academy 182–183
regenerative agriculture 13, 18, 19, 24, 48, 155, 189, 211
regional 30, 50, 51, 59, 90, 99, 102, 111, 113, 189, 192, 197, 204, 207–209, 211
Revel Meat... 162, 171, 173–175, 178, 179, 186, 187, 205
Rogers, Jeff 156, 162–166, 199, 122
roots 13, 15, 20, 24, 27, 122
ruminant 7, 21, 23, 25, 28, 31, 37, 48, 116, 117, 120, 122, 213, 214

S

sausage 10, 69, 74, 75, 77, 81–90, 94, 96–98, 101, 103, 109–112, 129, 136, 140, 142, 153, 157, 159, 168–170, 175, 195, 201–204, 209
Seattle 52, 145, 156, 157, 162, 166, 185
SFI 34, 40, 69, 214–218
sheep 21, 23, 28, 38, 41, 45, 88–91, 117, 122, 134, 138, 145, 146, 156, 162–164, 173, 175, 176, 199, 204, 213, 215, 217
shooting, shot 49, 50, 55, 59, 60, 70, 129, 148, 200
Siegel, Herbert 55, 60, 61
Sinclair, Upton 126, 127, 129
Skagit Valley . 145, 155, 163, 164
slaughter 0
- box 40, 57, 58, 60, 67, 128-129, 134, 200
- facility 51, 73, 76, 80, 90, 96, 112, 127-128, 132,

135-136, 140, 154, 163, 182, 198-199
- humane 11, 46-47, 49-51, 71-72, 89, 129, 178, 200, 203, 207, 212, 222
- mobile 10, 51, 58-59, 66-67, 69-71, 95-96, 106, 108, 114-115, 130, 133, 145, 148, 155, 156, 162, 170, 187, 193, 197
- MSU 52, 58, 67, 72, 96, 109, 110, 132–135, 137, 139, 141, 143–154, 163, 164, 170, 171, 194, 197–199, 204–206, 222
- on-farm .... 7, 11, 50-52, 57-60, 62, 65-67, 70 -71, 90, 96, 105, 113-114, 150-151, 193-194, 197, 199, 201, 204-205, 209
- regulation 43, 50, 52, 57, 99, 131, 200
- rifle 58-59, 62, 70, 90, 146, 200
- slaughtermen 10, 11, 44, 49, 66, 72, 138
- waste 100, 133, 137, 148
smoking 96, 102, 134
soil 7, 12,
soil erosion 12
soil quality . 7, 14, 15, 18, 19, 24, 33, 34, 36, 117, 219
Steinhaus, Klaus .. 74–78, 80–85
stockmanship 64, 65
Stockstill, Travis 151, 153–156, 194
Straubinger, Josef. 105–111, 113
stun 114, 115, 128, 141, 145, 147, 165, 166, 176–178, 200, 203
supply 6, 13, 39, 45, 52, 53, 84, 86, 88, 104, 131, 170,

173, 189, 192, 204, 205, 208, 209, 211
Sustainable Food Trust, SFT... 6, 12, 38, 42, 44, 207

T

Tönnies 99, 100, 104
trailer 19, 41, 47, 49, 55, 87, 89, 90, 95–97, 108, 115, 117, 133, 151, 164, 165, 178, 179, 192, 197–199
training............ 19, 51, 52, 66, 85, 89, 95, 101, 117
transport.............. 11, 35, 41, 49, 57–60, 69, 80, 89, 95, 96, 120, 133, 135, 163, 192, 193, 197–200
Tyson 133, 138, 139

U

Underly, Kari 183, 184
Uria MSB® 67
USDA 25, 35, 125–131, 133, 138, 145, 148, 150, 151, 155, 162, 170, 173, 182, 186

V

value adding 151, 194, 195
vegan 7, 23, 34–36, 172, 173, 178
vet 25, 43, 45, 50, 55, 57, 58, 60, 62, 69, 70, 71, 74, 75, 77, 80, 81, 85, 89, 90, 94–97, 110, 114, 115, 128, 133, 137, 141, 146, 155, 177, 178, 197, 198, 200, 204, 213
Visconti 157, 158
Vogel, Sarah 143, 144

W

wages........... 45, 85, 99, 103, 104, 155, 179, 186, 187, 206
Waizenegger, Rainer..... 73, 74, 84–86
Wallace, Nick 188–190, 192, 193, 207
Walter Smith Fine Foods ... 207, 208
Washington 145, 155, 181, 182, 184–186, 194, 199
waste 22, 75, 96, 100, 109, 111, 133, 137, 146, 148, 169, 198
wastewater 109, 146
water.............. 12, 14, 16, 17, 19, 22, 24, 25, 29, 36, 41, 58, 59, 70, 74, 76, 87, 96, 107, 109, 119, 120, 131, 133, 134, 139, 145, 146, 148, 149, 152, 164, 165, 178, 198, 213, 221
water infiltration 16, 17
Wenatchee........ 154, 157, 158, 159, 161
Wendle, Ralf 95, 96, 97, 101, 102, 104, 105, 129
Wilcox, Troy............ 185, 186
Wolfarth, Judith 91, 94–99, 101, 105, 110, 201
worms 12, 17, 19, 35

Y

yoke panel 68, 69, 108
Yellowstone 35

Z

Zarling, David 154–162, 166, 174, 186–188, 194, 195, 206

Gleaning a field after harvest.
Cattle play a vital role in
arable agriculture, too

Biographies

Marianne Landzettel is a journalist, writing and blogging about food, farming and agricultural policies in the UK, Germany and the US. She was UK and Ireland correspondent for German Public Radio, where she started her career as a reporter for the farming program. From 2003 until 2013 she worked for the BBC World Service. She is an avid supporter of organic and regenerative farmers, local processing and locally produced food and keen to promote knowledge exchange and agricultural practices that enhance biodiversity and soil health, in particular well-managed grazing.

Marianne is the author of: Regenerative Agriculture: Farming with Benefits. Profitable Farms. Healthy Food. Greener Planet.

X: @M_Landzettel
Internet: LondonCowGirl.com

Martin Kunz is a political scientist by training. He has been working in Fair Trade for almost 50 years, setting up Fair Trade supply chains as well as Fairtrade labeling associations. He enjoys taking photographs and has a keen interest in bees – his own as well as threatened Asian honey bee species (www.diversityhoneys.com). Last but not least he also is Marianne's husband. Apart from taking the pictures to go with her writing he deals with travel logistics – from transport to accommodation, including (some occasionally rather exotic) sleeping arrangements..

www.ingramcontent.com/pod-product-compliance
Lightning Source LLC
LaVergne TN
LVHW070946180726
843512LV00013B/954